Advances in Intelligent Systems and Computing

Volume 960

The series "Advances in Intelligent Systems and Computing" contains publications on theory, applications, and design methods of Intelligent Systems and Intelligent Computing. Virtually all disciplines such as engineering, natural sciences, computer and information science, ICT, economics, business, e-commerce, environment, healthcare, life science are covered. The list of topics spans all the areas of modern intelligent systems and computing such as: computational intelligence, soft computing including neural networks, fuzzy systems, evolutionary computing and the fusion of these paradigms, social intelligence, ambient intelligence, computational neuroscience, artificial life, virtual worlds and society, cognitive science and systems, Perception and Vision, DNA and immune based systems, self-organizing and adaptive systems, e-Learning and teaching, human-centered and human-centric computing, recommender systems, intelligent control, robotics and mechatronics including human-machine teaming, knowledge-based paradigms, learning paradigms, machine ethics, intelligent data analysis, knowledge management, intelligent agents, intelligent decision making and support, intelligent network security, trust management, interactive entertainment, Web intelligence and multimedia.

The publications within "Advances in Intelligent Systems and Computing" are primarily proceedings of important conferences, symposia and congresses. They cover significant recent developments in the field, both of a foundational and applicable character. An important characteristic feature of the series is the short publication time and world-wide distribution. This permits a rapid and broad dissemination of research results.

**** Indexing: The books of this series are submitted to ISI Proceedings, EI-Compendex, DBLP, SCOPUS, Google Scholar and Springerlink ****

More information about this series at http://www.springer.com/series/11156

Tareq Ahram · Waldemar Karwowski
Editors

Advances in Human Factors in Cybersecurity

Proceedings of the AHFE 2019 International Conference on Human Factors in Cybersecurity, July 24–28, 2019, Washington D.C., USA

Editors
Tareq Ahram
Institute for Advanced Systems Engineering
University of Central Florida
Orlando, FL, USA

Waldemar Karwowski
University of Central Florida
Orlando, FL, USA

ISSN 2194-5357 ISSN 2194-5365 (electronic)
Advances in Intelligent Systems and Computing
ISBN 978-3-030-20487-7 ISBN 978-3-030-20488-4 (eBook)
https://doi.org/10.1007/978-3-030-20488-4

This Springer imprint is published by the registered company Springer Nature Switzerland AG
The registered company address is: Gewerbestrasse 11, 6330 Cham, Switzerland

Advances in Human Factors and Ergonomics 2019

AHFE 2019 Series Editors

Tareq Ahram, Florida, USA
Waldemar Karwowski, Florida, USA

10th International Conference on Applied Human Factors and Ergonomics and the Affiliated Conferences

Proceedings of the AHFE 2019 International Conference on Human Factors in Cybersecurity, held on July 24–28, 2019, in Washington D.C., USA

Advances in Affective and Pleasurable Design	Shuichi Fukuda
Advances in Neuroergonomics and Cognitive Engineering	Hasan Ayaz
Advances in Design for Inclusion	Giuseppe Di Bucchianico
Advances in Ergonomics in Design	Francisco Rebelo and Marcelo M. Soares
Advances in Human Error, Reliability, Resilience, and Performance	Ronald L. Boring
Advances in Human Factors and Ergonomics in Healthcare and Medical Devices	Nancy J. Lightner and Jay Kalra
Advances in Human Factors and Simulation	Daniel N. Cassenti
Advances in Human Factors and Systems Interaction	Isabel L. Nunes
Advances in Human Factors in Cybersecurity	Tareq Ahram and Waldemar Karwowski
Advances in Human Factors, Business Management and Leadership	Jussi Ilari Kantola and Salman Nazir
Advances in Human Factors in Robots and Unmanned Systems	Jessie Chen
Advances in Human Factors in Training, Education, and Learning Sciences	Waldemar Karwowski, Tareq Ahram and Salman Nazir
Advances in Human Factors of Transportation	Neville Stanton

(continued)

(continued)

Advances in Artificial Intelligence, Software and Systems Engineering	Tareq Ahram
Advances in Human Factors in Architecture, Sustainable Urban Planning and Infrastructure	Jerzy Charytonowicz and Christianne Falcão
Advances in Physical Ergonomics and Human Factors	Ravindra S. Goonetilleke and Waldemar Karwowski
Advances in Interdisciplinary Practice in Industrial Design	Cliff Sungsoo Shin
Advances in Safety Management and Human Factors	Pedro M. Arezes
Advances in Social and Occupational Ergonomics	Richard H. M. Goossens and Atsuo Murata
Advances in Manufacturing, Production Management and Process Control	Waldemar Karwowski, Stefan Trzcielinski and Beata Mrugalska
Advances in Usability and User Experience	Tareq Ahram and Christianne Falcão
Advances in Human Factors in Wearable Technologies and Game Design	Tareq Ahram
Advances in Human Factors in Communication of Design	Amic G. Ho
Advances in Additive Manufacturing, Modeling Systems and 3D Prototyping	Massimo Di Nicolantonio, Emilio Rossi and Thomas Alexander

Preface

Our daily life, economic vitality, and national security depend on a stable, safe, and resilient cyberspace. We rely on this vast array of networks to communicate and travel, power our homes, run our economy, and provide government services. Yet, cyber intrusions and attacks have increased dramatically over the last decade, exposing sensitive personal and business information, disrupting critical operations, and imposing high costs on the economy. The human factor at the core of cybersecurity provides greater insight into this issue and highlights human error and awareness as key factors, in addition to technical lapses, as the areas of greatest concern. This book focuses on the social, economic, and behavioral aspects of cyberspace, which are largely missing from the general discourse on cybersecurity. The human element at the core of cybersecurity is what makes cyberspace the complex, adaptive system that it is. An inclusive, multi-disciplinary, holistic approach that combines the technical and behavioral element is needed to enhance cybersecurity. Human factors also pervade the top cyber threats. Personnel management and cyber awareness are essential for achieving holistic cybersecurity.

This book will be of special value to a large variety of professionals, researchers, and students focusing on the human aspect of cyberspace, and for the effective evaluation of security measures, interfaces, user-centered design, and design for special populations, particularly the elderly. We hope this book is informative, but even more than that it is thought-provoking. We hope it inspires, leading the reader to contemplate other questions, applications, and potential solutions in creating safe and secure designs for all.

This book includes two main sections:

Section 1 Cybersecurity Applications and Privacy Research
Section 2 Awareness and Cyber-Physical Security

Each section contains research papers that have been reviewed by members of the International Editorial Board. Our sincere thanks and appreciation to the Board members as listed below:

July 2019

Tareq Ahram
Waldemar Karwowski

Contents

Cybersecurity Applications and Privacy Research

Attempting to Reduce Susceptibility to Fraudulent Computer Pop-Ups Using Malevolence Cue Identification Training

Phillip L. Morgan[1,2(✉)], Robinson Soteriou[1,2], Craig Williams[1,2], and Qiyuan Zhang[1,2]

[1] Cognitive Science and Human Factors, School of Psychology, Cardiff University, Cardiff CF10 3AT, UK
{morganphil, soteriourm, williamscl31, zhangq47}@cardiff.ac.uk

[2] Cyber Psychology and Human Factors, Airbus Central R&T, The Quadrant, Celtic Springs Business Park, Newport NP10 8FZ, UK

Abstract. People accept a high number of computer pop-ups containing cues that indicate malevolence when they occur as interrupting tasks during a cognitively demanding memory-based task [1, 2], with younger adults spending only 5.5–6-s before making an accept or decline decision [2]. These findings may be explained by at least three factors: pressure to return to the suspended task to minimize forgetting; adopting non-cognitively demanding inspection strategies; and, having low levels of suspicion [3]. Consequences of such behavior could be potentially catastrophic for individuals and organizations (e.g., in the event of a successful cyber breach), and thus it is crucial to develop effective interventions to reduce susceptibility. The current experiment (N = 50) tested the effectiveness of *malevolence cue identification training* (MCIT) interventions. During phase 1, participants performed a serial recall task with some trials interrupted by pop-up messages with accept or cancel options that either contained cues (e.g., missing company name, misspelt word) to malevolence (*malevolent condition*) or no cues (*non-malevolent condition*). In phase 2, participants were allocated to one of three groups: no MCIT/Control, non-incentivized MCIT/N-IMCIT, or incentivized MCIT/IMCIT. Control group participants only had to identify category-related words (e.g., colors). Participants in intervention conditions were explicitly made aware of the malevolence cues in Phase 1 pop-ups before performing trying to identify malevolence cues within adapted passages of text. The N-IMCIT group were told that their detection accuracy was being ranked against other participants, to induce social comparison. Phase 3 was similar to phase 1, although 50% of malevolent pop-ups contained new cues. MCIT did lead to a significant reduction in the number of malevolent pop-ups accepted under some conditions. Incentivized training did not (statistically) improve performance compared to non-incentivized training. Cue novelty had no effect. Ways of further improving the MCIT training protocol used, as well as theoretical implications, are discussed.

Keywords: Cyber-security · Susceptibility · Task interruption · Intervention training

T. Ahram and W. Karwowski (Eds.): AHFE 2019, AISC 960, pp. 3–15, 2020.
https://doi.org/10.1007/978-3-030-20488-4_1

1 Introduction

The prevalence of malevolent online communications (MOCs), such as phishing attempts, is growing at a rapid pace. Recent statistics indicate that 264,483 phishing reports were made in the third quarter of 2018, which is markedly higher than the same quarter in 2016 [4]. The UK Government commissioned a report with the research revealing a staggering 46% of UK businesses reporting a breach of cyber-security, including phishing attempts, in the 12-months prior to being surveyed [5]. Such MOCs are targeted at individuals and organizations. Examples include fake pop-ups claiming to be from well-known companies that if clicked/accepted result in malware infection and/or payment demands [6]. Recent large-scale disruptive attacks include Sony Pictures, 2015, where employees clicked fake links resulting in login details & passwords being stolen, allowing fraudsters to hack-in [7]. The significance of this problem, together with other cyber threats, has been reflected by worldwide investments in cybersecurity with the likes of the UK Government and Bank of America committing funds to improve cybersecurity prevention and protection [8, 9]. Whilst much of this investment is being dedicated to improvements in the protection of networks, systems and software, computer users are seen as the main weakness in effective prevention of successful cyber-attack breaches [10], due to multiple fallibilities related to e.g., perception, attention, memory, decision making, and risk. Many cyber hackers are aware of these and will exploit them when developing MOCs. The current paper examines (1) susceptibility to MOCs delivered when humans are under short-term memory pressure and (2) the efficacy of an intervention training method to reduce susceptibility.

Pop-up messages occur regularly on networked computer devices, often unexpectedly and during engagement in another task(s) [11, 12]. Many contain information on and/or links to updates that are essential to maintain efficient performance of the computer system and/or software [13]. However, there are growing numbers of fake computer updates that mimic trusted companies and brands, with hackers' intent on encouraging people to clink on links which can result in cyber breaches.

Pop-ups at times will act as a distractor (e.g., if the user is able to ignore or deal with it without disengaging from an ongoing task) but are more likely to initiate an interruption (e.g., if the user is not able to ignore it and has to disengage from the ongoing task). Even short interruptions (as short as 2.8-s) can shift the focus of attention and memory and lead to increased errors within a suspended task [14] with factors such as interruption duration and demand exacerbating the extent of disruption [15, 16], as predicted by a leading model [17, 18]. However, few have considered how individuals choose to engage with interrupting tasks when there is no time constraint on their completion, i.e., when their response (which could be a few seconds) to the interrupting task determines when they will resume the suspended task (see [1, 2]).

In considering pop-up messages as task interruptions, how individuals allocate resources to verify authenticity will likely depend on factors outside the usual task parameters often studied, such as time costs and possible performance impairments. According to the Suspicion, Cognition, Automaticity Model/SCAM [3], whether malevolent cues are noticed within fraudulent communications depends on the depth of processing an individual engages in. The less suspicious and more trusting an individual is, the more likely they are to process the content of pop-up messages using automatic heuristic processing compared to someone who is more suspicious and less trusting who will likely engage in more cognitively effortful and time-consuming processing. Similarly, those who have a higher need for cognitive stimulation [19], will be more susceptible to influence techniques used within pop-up messages such as urgency, compliance with authority and avoidance of loss; at the expense of looking for suspicious aspects, such as message authenticity cues (e.g., correct spelling and grammar, company name). This leads to a prediction that an intervention training protocol that increases suspicion and encourages more effortful processing of pop-up message content should have carryover effects to a subsequent task performed with malevolent pop-up interruptions.

To our knowledge, only two published studies have considered human susceptibility to fraudulent pop-up interruptions occurring during a demanding memory-based task. [2] developed a paradigm where young adult participants were interrupted by one of three different types of pop-up message during a serial recall memory recall task. One third of pop-ups were designed to look genuine (*genuine* condition) and high in authority with no cues to potential malevolence. Another third (*mimicked* condition) were also high in authority but contained cues to suggest malevolence. The other third were also of a malevolent nature and *low authority* (i.e., contained no authority details relating to the source of the pop-up such as company name, logo, or website link). Participants had to decide whether to accept or decline pop-ups, at which point the primary task would be reinstated at the point of interruption. Predictions informed by parameters of SCAM [3] were supported, with an alarming 63% of mimicked pop-ups accepted compared with 66% in the genuine condition. Even more worrying was that 56% of low authority pop-ups were accepted. Participants spent on average only ~5.5–6-s viewing pop-up message content before committing to a response. When there were no time constraints to resume an interrupted task, participants accepted a slightly higher percentage (72%) of genuine pop-ups and slightly fewer (55%) mimicked pop-ups. This suggests that even without other cognitive and time pressures, people are still not very good at detecting malevolent cues within mimicked pop-up interruptions. [1] reported similar findings with older adults. Participants demonstrated higher levels of susceptibility to malevolent pop-ups during an interrupted memory recall phase, despite spending significant more time (~10.5–11-s) viewing them than in [1]. Fitting with SCAM-based low suspicion and automaticity predictions [3], both studies demonstrate very high levels of human susceptibility to malevolent pop-up interruptions that occur during a demanding memory-based task. However, concerns remain as neither study showed marked malevolent detection improvements when time pressure was not a factor.

Given these results, it is important further develop and test interventions to reduce susceptibility to computer-based communications such as malevolent pop-up messages. Education-based training interventions are not always effective [20] with some finding that people are more suspicious of scams that they are familiar with versus those that are less familiar [21]. [22] tested the effectiveness of emails containing cues to malevolence although found that not all people read and processed the content to a deep enough level to identify them effectively. These findings fit SCAM parameters regarding the use of automatic heuristic processing strategies, especially when suspicion is low. It could be that training effectiveness is dependent on the extent of encouragement to engage in and cognitively process training materials. Another factor that could potentially increase engagement in training is competition, which has been shown to facilitate motivation and performance in some instances [23]. Short term improvements were found when testing a competitive e-learning interface that displayed a rank order for the best performing students [24]. Competitive ranking may encourage individuals to engage more in the task to gain an accurate appraisal of their own performance compared to others, and thus improve upon such performances. The social process of evaluating one's accuracy in relation to their ability may encourage a desire to improve performance to increase own sense of self-worth [25]. Thinking more about the content of the training, as well as gaining satisfaction from it, may increase the likelihood of the information being remembered and utilized subsequently.

The current experiment has four main aims. One is to attempt to replicate the findings of [1] and [2] on susceptibility to pop-ups with cues to malevolence when they occur as interruptions to a memory-based task. Second, to examine whether, and if so to what extent, susceptibility can be alleviated through an intervention involving malevolent cue identification training (abbreviated to MCIT hereafter). The training was designed not to only increase suspicion and awareness of cues to malevolence but also to encourage more effortful cognitive processing of message content. Third, we examined whether a form of incentivized MCIT that encourages competitiveness through social comparison might further increase the intervention effectiveness. A fourth aim was to establish whether beneficial effects of MCIT transfer to conditions involving novel cues to malevolence that have not been experienced as part of the training intervention.

2 Method

2.1 Participants

Fifty Cardiff University Psychology undergraduate students (age: 19.32; *SD* 1.06) were recruited, via opportunity sampling, in return for course credits with adequate a priori power (.8 detect medium to large effect sizes (Cohen's f .25 −.4). Participants were first-language English or highly proficient in English as a second language and had normal/correct vision. They were assigned to one of three cue identification training groups. There were 16 in the Non-Malevolent Cue Identification (N-MCIT)/Control group (*M* age: 19.63-years, four male), 17 in the Non-Incentivized Malevolent Cue Identification (N-IMCIT) group (*M* age: 19.06-years, six male), and 17 in the Incentivized Malevolent Cue Identification (IMCIT) group (*M* age: 19.29-years, two male).

2.2 Design

A mixed factorial design was employed. The between-participants' independent variable (IV) was CIT Group with three levels: Control, N-IMCIT, and IMCIT. There were three repeated measures IVs. One was serial recall phase with two levels: Phase 1/Pre-Intervention 1, and, Phase 3/Post-Intervention. Another was the malevolency (Message

Type) of the pop-up with two levels: Non-Malevolent/Genuine, and, Non-Genuine/Malevolent. The third (Phase 3 only) was whether malevolent pop-ups contained the same (Malevolent-Old) or different malevolence cues than in Phase 1. There were two main dependent variables (DVs). The first was decision response to the pop-up request where two responses were possible: Accept, or, Decline. The second was the time to make a response. During the intervention phase, participant performance was recorded in two stages. The first stage required participants to respond to whether they identified at least one cue to indicate a category exemplar (Control group) or cue to malevolence (other groups), by choosing Yes or No. If choosing Yes, participants then had to record the number of cues identified (maximum 3 per passage of text with five passages in total).

2.3 Materials

Phase 1 and 3 Serial Recall and Interruption Pop-Up Tasks

Tasks were programmed on run on *Intel® Core™* i5 PCs connected to 1920 × 1080 *iiyama* 24″ flat-panel monitors. The serial recall task was created using PsychoPy2 software [26]. There were 18 trials in Phase 1 and 30 in Phase 3. During each trial, a different string of nine letters and numbers, e.g., 96KJ3785H were presented in the center of the screen for 9-s before disappearing. An instruction ('enter code') appeared after a 2-s retention interval to inform participants that they should try and recall and write down letters and numbers in the order in which they were presented.

Twelve trials were interrupted in Phase 1: six with non-malevolent and six with malevolent pop-ups. Six trials were not interrupted. Twenty-four trials were interrupted in Phase 3: twelve with non-malevolent and twelve with malevolent pop-ups, with six of these containing the same (Old) malevolency cues as in Phase 1 and six containing New cues. Pop-up messages appeared in the center of the screen after the letter/number string had disappeared and before the recall instruction appeared and remained on the screen until either an accept ('A' key) or cancel ('C') response was registered. Immediately after this response, the serial recall task was reinstated from the point in which it had been suspended (i.e., 'enter code' would appear next). Each new trial was initiated after the spacebar was pressed. Each pop-up contained text describing the scenario, plus an extra line of text with an instruction (e.g., 'Click 'accept' to download the [XXXX: name] pop -p') with boxes for Accept and Cancel. All non-malevolent and some malevolent pop-ups also contained a company logo in the top right corner and a hyperlink to a related website underneath text that read 'Further information can be found here:'.

Non-malevolent pop-ups contained cues (or indeed not lack of) to suggest that they were genuine (Fig. 1, left). These included a company logo, name (corresponding to logo), and website link, and accurate grammar and accurate spelling. Malevolent pop-ups (Fig. 1, right) contained three of six cues to malevolence: lack of company logo, name, website link, and an instance of inaccurate grammar or a misspelt word(s). During Phase 3, malevolent pop-ups contained either three Old or three New cues. New cues included: misspelling within website link, non-capitalization of company names, missing a key detail, having a fake logo, or capitalization of a word that should not be.

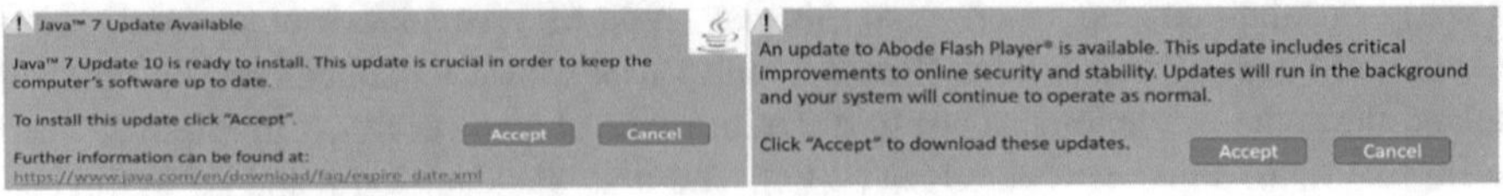

Fig. 1. Examples of a non-malevolent pop-up (left) and malevolent pop-up (right)

Prior to the start of Phase 1 and 3 trials, the following message was displayed in the middle of the computer screen for 15-s:

> *'This system is protected by virus protection software and pop-ups are installed on a regular basis. However, please be vigilant about the security of this system by ensuring that any attempts by applications to access system information of data are legitimate.'*

Phase 2 Intervention and Control Non-Intervention Tasks
Participants in the intervention conditions were given explicit information on the errors/cues to malevolence contained in Phase 1 malevolent pop-ups. They were also given a small whiteboard and marker pen to make notes on these if they wished to do so. All participants were required to read a set of five passages of text, with 5-min (~60-s per passage) from fictitious companies. The passages each contained textual information relating to five nominal categories (drinks, transport, sport, clothes, color). Passages were adapted for the N-IMCIT and IMCIT conditions to contain the same errors/cues to malevolence as in Phase 1. Participants in the Control group were required to first indicate whether category (e.g., color) words were present within the passage by clicking 'yes' or 'no' within an online answer sheet, and if choosing Yes, they then had to type the number of instances they could find (max = three per passage) before moving to the next passage. Participants in the intervention groups had to do this for cues indicating malevolence (max = 3 per passage) rather than category instances. Answer sheets were set out as a separate tab containing a table to be completed in relation to each passage. For the IMCIT group, each tab was followed by a leader board with performance *appearing* to be ranked against all previous participants with their position increasing after completion of each passage. Leaderboard positions were preset with the intention of encouraging (through social comparison) participants to try harder and apply more cognitive effort for each new passage.

2.4 Procedure

Before providing consent, participants read through an information sheet and experimental instructions (which were also verbally read by the experimenter) before completing two practice trials: one with a non-interrupted serial recall task, and another with a serial recall task interrupted by a non-malevolent pop-up. They were not informed about the cyber security element of the experiment during this process. At the beginning of Phase 1, participants were presented with the computer security message (see Materials). After this disappeared, they were able to press the spacebar to start trial one of 18, with 12 of the trials interrupted (see Materials). Phase 2 was the intervention phase. Participants read an instruction sheet appropriate for their group. All were instructed they had 5-min to read 5-passages (one-at-a-time) and complete the cue identification task relevant to their group. The Control group had to indicate (Yes or No) whether the passage of text contained at least one cue relating to its category description (e.g., color: look for color words). If answering yes, they then had to indicate how many category words they could identify within the passage (i.e., 1–3). N-IMCIT and IMCIT groups were first given written information pertaining to the malevolency cues contained within pop-ups experienced in Phase 1. These were explained verbally by the experimenter who checked participants' understanding. As with the Control group, participants in the MCIT groups were then presented with 5-passages of text, one-at-a-time, and had to indicate (Yes or No) whether the passage it contained at least one trained cue indicating potential malevolence. Participants were also provided with a small whiteboard and marker to make notes, if desired. Phase 3 (post-intervention) involved 30 serial recall trials with 24 interrupted. After Phase 3, participants completed demographics and pop-up awareness questionnaires. Participants were debriefed, with information about cyber-security and awareness aims.

3 Results and Discussion

All analyses are two-tailed with $\alpha = .05$. One dataset was excluded, as it was found to be a statistical outlier (z-scores > 3.29, $ps < .001$) on more than one measure.

Percentage of Pop-Up Messages Accepted/Declined
First, we consider mean percentages of '*malevolent*' pop-ups accepted across Phases 1 (pre-intervention) and 3 (post-intervention), collapsing across New and Old cue malevolent pop-ups in Phase 3 (Table 1). The percentage of malevolent pop-ups accepted looks to have decreased in Phase 3 for both MCIT groups, although increased for the Control group. Somewhat surprisingly, the mean percentage is markedly lower in the Control versus the N-IMCIT and IMCIT groups.

Table 1. Percentage of Malevolent and Genuine pop-ups accepted during Phases 1 and 2 and across each Training Group. *Note.* SD = Standard Deviation.

		Malevolent Pop-Ups		Genuine Pop-Ups	
Phase	Condition	Mean	SD	Mean	SD
1	Control	56.30	.34	63.54	.39
	N-IMCIT	73.41	.31	84.31	.30
	IMCIT	81.29	.29	87.25	.29
3	Control	67.69	.33	70.83	.35
	N-IMCIT	60.35	.31	85.29	.24
	IMCIT	70.12	.31	92.65	.11

A mixed 3 × 2 analysis of variance (ANOVA) with Training Group as the between-subjects variable (Control, N-IMCIT, IMCIT) and Phase (pre-intervention, post-intervention) revealed non-significant main effects of Training Group, $F(2, 47) = 1.07$, $MSE = .08$, $p = .35$, and, Phase, $F(1, 47) = 1.03$, $MSE = .04$, $p = .32$. There was however a significant interaction, $F(2, 47) = 3.44$, $MSE = .04$, $p = .04$. Bonferroni pot-hoc tests revealed a non-significant (although trend) reduction in the percentage of malevolent pop-ups accepted in Phase 3 compared with Phase 1 for the IMCIT group ($p = .07$). However, the significant interaction might be better explained by the percentage of malevolent pop-ups accepted by the Control group in Phase 1 being significantly lower than in the N-IMCIT and IMCIT groups within Phase 1 ($ps < .025$). Given this unexpected difference (discussed later), another mixed ANOVA, this time 2 (Training Group: MCIT, IMCIT) × 2 (Phase: 1, 3), was conducted. This revealed a significant main effect of Phase, $F(1, 32) = 5.63$, $MSE = .04$, $p = .02$ with a lower percentage of malevolent pop-ups accepted in Phase 3 than in Phase 1. There was a non-significant main effect of Training Group, $F(1, 32) = .96$, $MSE = .08$, $p = .33$, and a non-significant interaction, $F(1, 32) = .03$, $MSE = .04$, $p = .86$.

Taken together, these findings suggest that: (1) MCIT worked in terms of reducing the percentage of malevolent pop-up messages accepted post-intervention, (2) IMCIT did not lead to better performance than N-IMCIT, and, (3) participants in the Control group, in Phase 1 at least, performed differently (i.e., chose to accept far less malevolent pop-ups) to those in MCIT conditions. In relation to (1), findings are in line with SCAM predictions that heightening suspicion will lead to increased cognitive and less automatic processing of stimuli [3], thus improving the likelihood of identifying malevolence cues. However, the percentage of malevolent pop-ups accepted was still very high, even after the intervention. In relation to (2), incentivized MCIT through social comparison (using an onscreen leaderboard technique), was not effective enough to cause even more suspicion and increased cognitive processing of potential cues to suggest malevolence within pop-up messages compared to non-incentivized MCIT. This finding (despite there being a trend) is not in line with [22] and possible reasons are considered in the Limitations section. Considering (3), the only difference was when the groups were tested: The Control group were tested after the MCIT groups.

Next, we consider mean percentages of *'genuine'* pop-ups accepted in Phases 1 and 3, noting again that both New and Old cue malevolent pop-up data are collapsed across (Table 1). The percentage of genuine pop-ups accepted increased marginally in Phase 3 across all groups. However, and as with malevolent pop-ups, the mean percentage of genuine pop-ups accepted in Phase 1 was markedly lower in the Control versus MCIT groups. A mixed 3 × 2 analysis of variance (ANOVA) with Training Group as the between-subjects variable and Phase revealed a marginally non-significant main effect of Training Group, $F(2, 47) = 3.12$, $MSE = .07$, $p = .054$, and a non-significant main effect of Phase, $F(1, 47) = 2.57$, $MSE = .02$, $p = .12$. There was a non-significant interaction. However, these findings might again be affected by the unusual pattern of data in the Control condition during Phase 1 compared to the MCIT condition. Therefore, a 2 (Training Group: MCIT, IMCIT) × 2 (Phase: 1, 3) mixed ANOVA was conducted. There were non-significant main effects of Training Group, $F(1, 32) < 1$, $p = .50$, and Phase, $F(1, 32) < 1$, $p = .39$, and a non-significant interaction, $F(1, 32) < 1$, $p = .55$.

Taken together, these findings suggest that (1) the ability to identify genuine pop-up messages was high, (2) MCIT did not have any effect on this, and (3) participants in the Control group, in Phase 1 at least, performed quite differently (i.e., accepted fewer genuine pop-ups) to those in the MCIT conditions. It is difficult to determine why participants in MCIT groups seemed to be very good at classifying most genuine pop-ups as genuine and then chose to accept rather than decline. It might have been relatively easier to check whether pop-ups contained no cues to malevolence than to check and register a cue(s) to malevolence. Although, and given the very high (and somewhat worrying) percentages of malevolent pop-ups accepted, it could be that participants, particularly in Phase 1, were more inclined to adopt a trusting stance [3] and accept most pop-ups as being genuine unless they noted at least one cue that was enough to raise suspension and cause them to respond in a different way (i.e., decline the pop-up. In order to speak to these possibilities, we will later examine the amount of time participants took before making a decision to accept/decline messages.

Next, we examine for possible differences between the percentage of Old (i.e., contained same cue types as in Phase 1, trained on these cues in MCIT conditions in Phase 2) versus New (i.e., contained different cue types as in Phase 1, not trained on these cues in MCIT conditions in Phase 2) malevolent pop-ups in Phase 3 only (Table 2). Whilst there is no difference within the Control Group, participants in the MCIT groups appear to have accepted marginally more New than Old malevolent messages, particularly in the IMCIT condition. However, a 3 (Training Group) $\times$ 2 (Cue Familiarity: Old, New) mixed ANOVA revealed non-significant main effects of Training Group, $F(2, 47) < 1$, $p = .64$, Cue Familiarity, $F(1, 47) < 1$, $p = .92$, and a non-significant interaction, $F(2, 47) < 1$, $p = .33$. Given the unusual accept/decline behavior of the Control Group in Phase 1 (see above), an additional analysis (2 $\times$ 2 mixed ANOVA) was conducted with the Control group excluded. There were still non-significant main effects of Training Group, $F(1, 32) < 1$, $p = .36$, Cue Familiarity, $F(1, 32) = 1.61$, $p = .21$, and a non-significant interaction, $F(1, 32) < 1$, $p = .62$.

Table 2. Percentage of Old and New pop-ups accepted during Phase 3 across each Training Group. *Note.* SD = Standard Deviation.

		Malevolent Pop-Ups	
Phase	Condition	Mean	SD
Old	Control	67.75%	.34
	N-IMCIT	58.82%	.35
	IMCIT	66.65%	.34
New	Control	67.75%	.36
	N-IMCIT	61.79%	.31
	IMCIT	73.35%	.30

We anticipated that participants in both MCIT groups, and in particular the I-MCIT group would be less likely to spot new cues. However, there is no statistical evidence to suggest that any form of MCIT led to participants accepting more New messages, despite an ~11.5% higher acceptance of these in the IMCIT versus the N-IMCIT condition in Phase 3. Of course, this could be a power issue, and future studies should consider this before ruling out the possibility that MCIT will not put people at a disadvantage in terms of spotting malevolent cues that they have not be trained to identify,

Time to Accept/Decline Pop-Up Messages

Next, we consider the time taken at make an accept/decline response. Noting that the time to accept/decline malevolent pop-ups was 5.37-s for younger adults in the [2] study, and 10-92-s for older adults in the [1] study. In the same studies, the times to accept genuine pop-ups were 5.47-s and 10.45-s respectively. Mean pop-up accept/decline times for the current study are displayed in Table 3 (with one outlier removed: z-scores > 3.29, $p < .001$). Malevolent and genuine pop-ups, accept/decline times are noticeably lower ($\sim$1–2-s) than in e.g., [2]. Also, response times appear to reduce for each Group in Phase 3 versus Phase 1. The third, and somewhat counter-intuitive observation, is that response times are noticeably lowest (and very short) for the Control Group (*M* 3.39 Phase 1, *M* 2.97 Phase 3).

Table 3. Time (seconds) before making an accept/decline response to Malevolent and Genuine pop-ups during Phases 1 and 2 and across each Training Group. *Note.* SD = Standard Deviation.

		Malevolent Pop-Ups		Genuine Pop-Ups	
Phase	Condition	Mean	SD	Mean	SD
1	Control	3.45	2.65	3.33	2.23
	N-IMCIT	4.41	2.49	4.26	2.15
	IMCIT	4.77	2.99	4.57	3.05
3	Control	3.08	1.82	2.85	1.87
	N-IMCIT	3.66	1.57	3.51	1.55
	IMCIT	4.54	3.07	4.05	2.31

A 3 (Training Group) × 2 (Phase) × 2 (Message Type) mixed factorial ANOVA revealed a significant main effect of Phase, $F(1, 46) = 4.55$, $MSE = 2.87$, $p = .038$, with less time taken in Phase 3 ($M = 3.62$-s) than Phase 1 (*M* 4.13). There was a significant main effect of Message Type, $F(1, 46) = 5.46$, $MSE = .45$, $p = .024$, with more time spent before making an accept/decline response for malevolent (*M* 3.99) than genuine (*M* 3.76-s) messages. There was a non-significant main effect of Training Group, $F(2, 46) = 1.55$, $MSE = 4.47$, $p = .22$, and none of the interactions were significant (*ps* > .08).

Contrary to our prediction, participants were faster to respond to pop-up messages in Phase 3 than Phase 1, and despite a non-significant Phase × Training Group interaction, this was the case for the IMCIT (*M Diff* −0.23-s) and N-IMCIT (*M Diff* −0.75-s) groups. Given that participants in the MCIT groups did not take additional time to try and identify cues to malevolence in malevolent pop-up messages, the improved detection performance must have been due to increased suspicion [3] and making better use of the very short inspection times to identify at least one cue to rouse suspicion.

Given much lower acceptance rates of malevolent pop-ups amongst the Control group in Phase 1 (Table 1), it was expected that those participants took more time to try and identify cues than in the MCIT groups. This was not the case. Also, their acceptance rate for malevolent pop-ups in Phase 3 *increased* by over 10% and the time taken to accept/decline messages reduced by almost half a second. Upon closer inspection of the data, three Control group participants almost always declined malevolent messages compared with the others whose performance was largely in line with those in the MCIT groups. However, they were not statistical outliers the $p < .001$ (z-scores > 3.29) level.

4 Limitations

There are limitations. First, there was no statistical evidence to suggest that those in the IMCIT group were better at identifying malevolent pop-ups than those in the N-IMCIT group, despite a trend. Perhaps using a leaderboard with individual position increasing after each task (e.g., $19^{th}/20$ after the first task, 1st after the last task) was not effective enough. This may be influenced by some participants potentially being aware that they were performing optimally and met with incongruent feedback to suggest otherwise. Competing with other people *in situ* may have promoted stronger social comparison and led to more intense cognitive processing strategies [3]. Second, within both MCIT conditions, participants had to identify whether they detected malevolent cues and then type a number corresponding to how many. This method meant that accuracy of malevolent cue identical could not be measured. Third, participants had one-minute per training task, only five tasks to complete, with each passage containing only three malevolent cues. They were also aware that there would be a maximum of three malevolent cues. This may not have been cognitively engaging enough. Finally, Control group participants were treating pop-ups with higher levels of suspicion in Phase 1. Ideally, this condition would be re-run to check for a possibly anomalous effect.

5 Implications

We successfully demonstrated that MCIT can be used as an intervention to reduce susceptibility to potentially fraudulent computer pop-ups. More cognitively engaging and demanding versions of the intervention might be even more effective. Whilst an incentivized version of this intervention did not quite result in further improvements in identifying fraudulent pop-ups, an improved version that better encourages social comparison might work better. Whilst there was no statistical evidence to suggest that MCIT can impair the ability to detect malevolence cues that participants had not been trained on, trends indicated a performance deficit, and methods to mitigate this need to be considered in the future development of MCIT interventions. Finally, it is important to note that time spent viewing malevolent pop-up messages was incredibly low and the propensity to accept (rather than decline) them was alarmingly high, both pre- and post- intervention, and even higher than in the studies by [1 and 2]. This further emphasises the vital need to develop interventions to help alleviate such susceptibility.

References

1. Morgan, P.L., Williams, E.J., Zook, N.A., Christopher, G.: Exploring older adult susceptibility to fraudulent computer pop-up interruptions. In: International Conference on Applied Human Factors and Ergonomics, pp. 56–68. Springer, Cham (2018)
2. Williams, E.J., Morgan, P.L., Joinson, A.N.: Press accept to update now: individual differences in susceptibility to malevolent interruptions. Decis. Support Syst. **96**, 119–129 (2017)
3. Vishwanath, A., Harrison, B., Ng, Y.J.: Suspicion, cognition, and automaticity model of phishing susceptibility. Commun. Res. **45**(8), 1–21 (2016)
4. Anti-Phishing Working Group (APWG). https://www.antiphishing.org/resources/apwg-reports/
5. Department for Culture, Media & Sport.: Cyber security breaches survey 2017. https://www.gov.uk/government/uploads/system/uploads/attachment_data/file/609186/Cyber_Security_Breaches_Survey_2017_main_report_PUBLIC.pdf
6. National Cyber Security Centre.: Weekly threat report, 30 June 2017. https://www.ncsc.gov.uk/report/weekly-threat-report-30th-june2017
7. Perera, D.: Researcher: Sony hackers used fake emails, Politico. https://www.politico.com/story/2015/04/sony-hackers-fake-emails-117200
8. Forbes Cyber Security report. https://www.forbes.com/sites/ellistalton/2018/04/23/the-u-s-governments-lack-of-cybersecurity-expertise-threatens-our-infrastructure/#20d248be49e0
9. HM Government. National cyber security strategy 2016–2021. https://www.gov.uk/government/uploads/system/uploads/attachment_data/file/567242/national_cyber_security_strategy_2016.pdf
10. Conteh, N.Y., Schmick, P.J.: Cybersecurity: risks, vulnerabilities and countermeasures to prevent social engineering attacks. Int. J. Adv. Comput. Res. **6**(23), 31 (2016)
11. Downing, D., Covington, M., Covington, M., Barrett, C.A., Covington, S.: Dictionary of Computer and Internet Terms. Barron's Educational Series, New York (2000)
12. Daintith, J., Wright, E.: A Dictionary of Computing. Oxford University Press, Oxford (2008)
13. Norton How To 2018. https://us.norton.com/internetsecurity-how-to-the-importance-of-general-software-updates-and-patches.html
14. Altmann, E.M., Trafton, J.G., Hambrick, D.Z.: Momentary interruptions can derail the train of thought. J. Exp. Psychol. Gen. **143**(1), 215–226 (2014)
15. Hodgetts, H.M., Jones, D.M.: Interruption of the Tower of London task: support for a goal-activation approach. J. Exp. Psychol. Gen. **135**(1), 103–115 (2006)
16. Monk, C.A., Trafton, J.G., Boehm-Davis, D.A.: The effect of interruption duration and demand on resuming suspended goals. J. Exp. Psychol. Appl. **14**(4), 299–313 (2008)
17. Altmann, E.M., Trafton, J.G.: Memory for goals: an activation-based model. Cogn. Sci. **26**, 39–83 (2002)
18. Altmann, E.M., Trafton, J.G.: Timecourse of recovery from task interruption: data and a model. Psychon. Bull. Rev. **14**(6), 1079–1084 (2017)
19. Cacioppo, J.T., Petty, R.E., Feng Kao, C.: The efficient assessment of need for cognition. J. Pers. Assess. **48**(3), 306–307 (1984)
20. Anandpara, V., Dingman, A., Jakobsson, M., Liu, D., Roinestad, H.: Phishing IQ tests measure fear, not ability. In: International Conference on Financial Cryptography and Data Security, pp. 362–366. Springer, Berlin (2007)
21. Downs, J.S., Holbrook, M.B., Cranor, L.F.: Decision strategies and susceptibility to phishing. In: Proceedings of the Second Symposium on Usable Privacy and Security, pp. 79–90. ACM (2006)

22. Kumaraguru, P., Sheng, S., Acquisti, A., Cranor, L.F., Hong, J.: Teaching Johnny not to fall for phish. ACM Trans. Internet Technol. (TOIT). **10**(2), 1–30 (2010)
23. Clifford, M.M.: Effects of competition as a motivational technique in the classroom. Am. Educ. Res. J. **9**(1), 123–137 (1972)
24. Aleman, J.L.F., de Gea, J.M.C., Mondéjar, J.J.R.: Effects of competitive computer-assisted learning versus conventional teaching methods on the acquisition and retention of knowledge in medical surgical nursing students. Nurse Educ. Today **31**(8), 866–871 (2011)
25. Festinger, L.: A theory of social comparison processes. Hum. Relat. **7**(2), 117–140 (1954)
26. Peirce, J.W.: PsychoPy—psychophysics software in Python. J. Neurosci. Methods **162**(2), 8–13 (2007)

Cyber Resilient Behavior: Integrating Human Behavioral Models and Resilience Engineering Capabilities into Cyber Security

Rick van der Kleij[1,2(✉)] and Rutger Leukfeldt[1,3]

[1] Cybersecurity and SMEs Research Group, The Hague University of Applied Sciences (THUAS), The Hague, The Netherlands
{R.vanderkleij,E.R.Leukfeldt}@hhs.nl
[2] Department of Human Behavior and Organisational Innovations, The Netherlands Organisation for Applied Scientific Research (TNO), The Hague, The Netherlands
[3] Netherlands Institute for the Study of Crime and Law Enforcement (NSCR), Amsterdam, The Netherlands

Abstract. Cybercrime is on the rise. With the ongoing digitization of our society, it is expected that, sooner or later, all organizations have to deal with cyberattacks; hence organizations need to be more cyber resilient. This paper presents a novel framework of cyber resilience, integrating models from resilience engineering and human behavior. Based on a pilot study with nearly 60 small and medium-sized enterprises (SMEs) in the Netherlands, this paper shows that the proposed framework holds the promise for better development of human aspects of cyber resilience within organizations. The framework provides organizations with diagnostic capability into how to better prepare for emerging cyber threats, while assuring the viability of human aspects of cyber security critical to their business continuity. Moreover, knowing the sources of behavior that predict cyber resiliency may help in the development of successful behavioral intervention programs.

Keywords: Security behaviors · Human aspects of cyber security · Cyber hygiene behavior

1 Introduction

Every day, companies all over the world suffer cyberattacks. The number of cyberattacks on organizations worldwide has been growing over the last six years to an average of 2.5 attacks every week for large organizations [1]. Based on 2,182 interviews from 254 companies in seven countries, the Ponemon institute [1] calculated that the average cost of cybercrime in 2017 was 11.7 million US dollars per organization. These costs are internal, dealing with cybercrime and lost business opportunities, and external, including the loss of information assets, business disruption, equipment damage and revenue loss. With the ongoing digitization of our society, it is expected that the number of cyberattacks and, consequently, the annual costs of cybercrime, will increase rapidly over the next years.

T. Ahram and W. Karwowski (Eds.): AHFE 2019, AISC 960, pp. 16–27, 2020.
https://doi.org/10.1007/978-3-030-20488-4_2

Organizations are often unprepared to face cyberattacks, to recover from attacks and lack formal incident response plans [2]. Only 32% of IT and security professionals say that their organization has a high level of *cyber resilience*, and this number is decreasing. A recent development is that security breaches are becoming more damaging to organizations, thus accelerating the need for cyber resilience [3]. As it is expected that the number of cyberattacks will also grow in the near future, the notion that organizations need to be cyber resilient is becoming increasingly popular [4]. This notion has been put forward in perspectives from multiple disciplines, including systems engineering, auditing and risk assessment. The general idea is that instead of focusing all efforts on keeping criminals out of company networks, it is better to assume they will eventually break through the organizations' defenses, and to start working on increasing cyber resilience to reduce the impact[1]. In order to ensure resilience, it is necessary to first accurately define and measure it [4]. There is, however, still a lot of confusion about what the term resilience means [5].

The notion that organizations need to be more cyber resilient is quite a recent one and, because most reports have been published by consultancy firms, commercial companies, and private research institutes working on behalf of these businesses, the scope of the scientific literature in this particular domain is fairly limited. The review of these few studies shows, however, that cyber resilience, and organizational resilience more broadly, is an emerging field [6]. Further, as pointed out by Parsons and colleagues [7], the study of this emerging field has predominantly been focused on the technical issues of cyber resilience. Consequently, measures to enhance cyber resiliency are mainly focused on improving the existing security infrastructure. It is only recently, with some exceptions, that researchers have started looking at the human aspects as potential sources of organizational resilience [8–10].

The human aspects as potential sources of organizational resilience should not be underestimated. It is a well-established fact that in many cyberattacks the behaviors of employees are exploited [11–13]. Regardless of how secure a system is, the end user is often a critical backdoor into the network [14–16]. Attackers look for vulnerabilities to gain access to systems; these can come from users who exhibit cyber risky behaviors, such as by not following best practices or revealing too much personal information on social networking sites. Hence, cyber resilience includes protecting against harm that may come due to malpractice by insiders, whether intentional or accidental. Furthermore, employees can perform many different behaviors to protect themselves and their organization from computer security threats [17]. Through good practice or cyber hygiene behavior, such as using strong passwords and responding adequately to incidents, employees may help the organization to become more cyber resilient (see also [9, 18–20]). Against this background, this paper aims to (1) develop a novel comprehensive and coherent framework of cyber resilience integrating models from resilience engineering and human behavior, and (2) to test this framework in a pilot study with 56 small and medium-sized enterprises (SMEs) in the Netherlands. And although this paper focuses on the human aspects of cyber security, it also acknowledges that technological factors play an important role in cyber security at the organizational level.

[1] https://www.itgovernance.co.uk/cyber-resilience.

This paper is structured as follows. First, we start by giving an overview of current views on resilience. Then we discuss abilities necessary for resilience at the organizational level and show that these abilities are relevant in the domain of cyber security as well. We then combine a framework for understanding human behavior with a resilience engineering model and discuss a combined model of resilient behavior. Next, we describe how the framework was operationalized into a questionnaire to measure resilient behavior within the domain of cyber security and present the results of a small empirical test of the framework and instrument. We conclude with a discussion of our work and give directions for future research.

2 Resilience

Research on resilience has increased substantially over the past decades, following dissatisfaction with traditional models of risk and vulnerabilities, which focus on the results of adverse events [21]. Whereas traditional risk models have historically been useful for many applications in the past, it is difficult to apply them to cyber risks [4, 22]. Traditional risk assessment approaches tend to break down when it is difficult to clearly identify the threats, assess vulnerabilities, and quantify consequences [23, 24]. Cyber threats cannot be clearly identified and quantified through historical measures due to the rapidly changing threat environment [4]. Resilience, however, focuses on the ability to succeed under varying conditions.

Resilience was first used in medical and material sciences, relating to the ability to recover from stress or strain [25]. More recently a wider concept of resilience has emerged. Within the domain of health sciences, the Resilience and Healthy Ageing Network defined resilience as "the process of negotiating, managing and adapting to significant sources of stress or trauma. Assets and resources within the individual, their life and their environment facilitate this capacity for adaptation and 'bouncing back' in the face of adversity. Across the life course, the experience of resilience will vary" ([21], p. 2).

Matzenberger [25] defines resilience in a learning environment as the capacity or ability of a system to persist after disturbance and to reorganize or emerge while sustaining essentially the same function. In hazards research, resilience is understood as the ability to survive and cope with a disaster with minimum impact and damage. It holds the capacity to reduce or avoid losses, contain effects of disasters, and recover with minimal disruptions [25, 26]. In resilience engineering, a new approach to risk management, resilience is described at the generic system level as the intrinsic ability of a system to adjust its functioning prior to, during, or following changes and disturbances, so that it can sustain required operations under both expected and unexpected conditions [27]. Van der Beek and Schraagen [28] have defined resilience at the team level in crisis management situations by expanding the ability of a system with more relation-oriented abilities of leadership and cooperation.

The key term to all these definitions is the system's ability to adjust its functioning. Resilience represents the capacity (of an organizational system) to anticipate and manage risk effectively, through appropriate adaptation of its actions, systems and processes, so as to ensure that its core functions are carried out in a stable and effective relationship with the environment [29]. Although these definitions at a glance mainly seem to differ in level of analysis, ranging from the individual via the team level to the more generic organizational (system) level, Woods [5] argues that resilience is used in four different ways: (1) resilience as rebound from trauma and return to equilibrium; (2) resilience as a synonym for robustness; (3) resilience as the opposite of brittleness, i.e., as graceful extensibility when surprise challenges boundaries; (4) resilience as network architectures that can sustain the ability to adapt to future surprises as conditions evolve. The implication of this partition is that one needs to be explicit about which of the four senses of resilience is meant when studying or modeling adaptive capacities (or to expand on the four anchor concepts as new results emerge) [5]. Not all these uses of the label 'resilience' are correct according to Woods. Resilience as the increased ability to absorb perturbations confounds the labels robustness and resilience. Some of the earliest explorations of resilience confounded these two labels, and this confound continues to add noise to work on resilience [5].

The broad working definitions of resilient performance can be made more precise and operational by considering what makes resilient performance possible. Since resilient performance is possible for most, if not all, systems, the explanation must refer to something that is independent of any specific domain. Hollnagel [30] has proposed the following four abilities necessary for resilient performance (see also [27]):

- The ability to *Anticipate*. Knowing what to expect or being able to anticipate developments further into the future, such as potential disruptions, novel demands or constraints, new opportunities, or changing operating conditions. This is the ability to create foresight and to address the potential.
- The ability to *Monitor*. Knowing what to look for or being able to monitor that which is or could seriously affect the system's performance in the near term, positively or negatively. The monitoring must cover the system's own performance as well as what happens in the environment. This is the ability to address the critical.
- The ability to *Respond*. Knowing what to do, or being able to respond to regular and irregular changes, disturbances, and opportunities by activating prepared actions or by adjusting current modes of functioning. It includes assessing the situation, knowing what to respond to, finding or deciding what to do, and when to do it. This is the ability to address the actual.
- The ability to *Learn*. Knowing what has happened, or being able to learn from experience, in particular to learn the right lessons from the right experience, successes as well as failures. This is the ability to address the factual. Although this capacity is often overlooked, it is a critical aspect of resilience. By learning how to be more adaptable, systems are better equipped to respond when faced with some sort of disruption [25].

Hence, the resilience of a system is defined by the abilities to respond to the actual, to monitor the critical, to anticipate the potential, and to learn from the factual [27]. The abovementioned abilities can be thought of together as a framework for identification and classification of indicators [25]. The engineering of resilience comprises the ways in which these four capabilities can be established and managed [27]. This is of importance to organizations because being resilient can provide them with a competitive advantage [31]. Resilient organizations may also contribute to a cyber-resilient community or to more cyber resiliency at the nationwide level. McManus [32] argues that resilient organizations contribute directly to the speed and success of community resilience. Without critical services provided by resilient organizations, such as transport, healthcare and electricity, communities (or states alike) cannot respond or recover (see also [33]).

Although these four capabilities have been under debate for quite some time, and research at the employee [34] and team level [28] only partially support the four-dimensional nature of the construct as proposed by Hollnagel [27], we feel that these four dimensions have high face validity in the domain of cyber security. For instance, the recently completed NIST framework for improving Critical Infrastructure Cybersecurity encompasses five similar functions [35]. These five functions are Identify, Protect, Detect, Respond, and Recover. They aid an organization in expressing its management of cybersecurity risk by organizing information, enabling risk management decisions, addressing threats, and improving by learning from previous activities. The functions also align with existing methodologies for incident management and help show the impact of investments in cybersecurity. For example, investments in planning and exercises support timely response and recovery actions, resulting in reduced impact to the delivery of services.

We feel that the four abilities as proposed by Hollnagel [27] together seem to be sufficient without being redundant. We see no need, for instance, to split the function Anticipate into the NIST functions of Identify and Protect. We think that the Protect function, in which appropriate safeguards are developed and implemented to ensure delivery of critical infrastructure services, is a composite rather than a primary ability of a resilient organization. Implementing appropriate safeguards is a combination of the ability to Anticipate and to Learn, and possibly also the ability to Detect (see also, [30]). Moreover, the Identify function has a strong focus on understanding the contexts and the risks, while Anticipate also looks at opportunities and emerging threats. We also feel that the Recover function, to develop and implement the appropriate activities to maintain plans for resilience and to restore any capabilities or services that were impaired due to a cybersecurity incident, is essential, but again, we do not think of this ability as a primary function. This function is the consequence of another ability, namely Learning. For sustainable recovery to take place within organizational systems, organizations should have knowledge of what has happened, or should be able to learn from experience.

3 Human Behavior

There is an important role for employees within organizations to help the organization become more cyber resilient. In principle, one could easily envisage employees performing Monitoring, Responding, Anticipating and Learning functions to maintain resilience in cyber capabilities. To explain the four resilience functions from a behavioral perspective, a comprehensive framework for understanding human behavior can be applied that involves three essential conditions: Motivation, Opportunity, and Ability (MOA) [36]. The cyber resiliency of an organization is in part determined by employees' motivation, opportunity, and ability to perform the four generic resilience functions (see also [36]). In a recent application of this framework, capability, opportunity, and motivation interact to generate behavior, such as responding to a cyber-security incident, that in turn influences these components (the 'COM-B' system, see [37]). "Capability is defined herein as the individual's psychological and physical capacity to engage in the activity concerned. It includes having the necessary knowledge and skills. Motivation is defined as all those brain processes that energize and direct behavior, not just goals and conscious decision-making. It includes habitual processes, emotional responding, as well as analytical decision-making. Opportunity is defined as all the factors that lie outside the individual that make the behavior possible or prompt it" ([37], p. 4). Examples of Opportunity are social support and organizational climate.

Although we do not oppose the idea that the capacity to be resilient depends on the technical infrastructure of an organization, we believe that a key component of being resilient and for assessing the resilient capacity of organizations resides, for a large part, at the individual employee level. Employees need to have the psychological and physical *abilities* to act in a cyber-resilient manner. Further, employees need to have the right mindset to do so. They need to be *motivated* to behave in a cyber-resilient manner. Finally, the organization needs to provide *opportunities* to enable the desired behavior. People can be capable and motivated to behave in a resilient manner, but when there is no opportunity to do so within the organization, for instance because resources are lacking (e.g., a backup and restore system), these intentions remain in vain.

An important reason for us to utilize the COM-B system to explain resilience functions from a behavioral perspective, is that this system is part of a larger framework of behavior change interventions. Based on a systematic review of literature, Michie and colleagues [37] identified nine generic intervention functions. The nine intervention functions are aimed at addressing deficits in one or more of the conditions: capability, opportunity, and motivation. These functions are explicitly linked to one of more of these conditions. An example is 'persuasion': Using communication to induce positive or negative feelings or stimulate action. Furthermore, seven policy categories were identified that could enable those interventions to occur. An example is 'regulation', or, in other words, establishing rules or principles of behavior or practice. This means that the framework proposed in Sect. 4 hereafter, although not explicitly embedded, holds the power to link intervention functions and policy categories to an analysis of the targeted cyber resilient behavior of employees within organizations. Thus, in the context of cyber security, the framework could serve the need for more 'fit for purpose interventions' to change human cyber behavior, as stipulated by Young and colleagues [13]. For instance, when lack of motivation hinders resilient functioning of an organization, a suitable intervention function could be to apply 'modelling': providing an example for people to aspire to or imitate. A policy category that could be used to enable modeling is 'marketing': Using print, electronic, telephonic or broadcast media. Just by identifying all the potential intervention functions and policy categories this behavior change framework could prevent policy makers and intervention designers from neglecting important options [37].

4 Conceptual Framework of Resilient Behavior

It is important to be able to measure cyber resilience. Metrics can contribute to key organizational needs, such as the need to demonstrate progress towards becoming more resilient, the need to demonstrate the effects of an intervention, or the need to link improvements in resilience with increased competitiveness [33]. As we have argued, resilience metrics could also contribute to the design of successful behavior change programs. Interventions could be tailored to the specific needs of the organization based on diagnosed sources of non-resilient behavior.

In this paper, we have combined the definitions of the four resilience functions and the three sources of behavior to create a conceptual framework of resilient behavior (see Table 1). We then tailored the model to cyber security by drawing upon metrics from the cyber security literature, primarily from [38, 39] and [9]. Further, because the aim is to measure employees' motivation, opportunity, and ability to perform the four generic resilience functions, the framework was operationalized into a questionnaire: The Cyber Resilient Behavior Questionnaire.

For each of the four resilience functions, we developed several specific capability statements, opportunity statements and motivation statements. For example, the following statements measure the function 'Respond':

Capability: "Employees in our organization know what to do in the event of cybercrime."

Opportunity: "In the event of cybercrime, our organization has clearly defined procedures on how to deal with disruptions to business operations."

Motivation: "Employees in our organization consider the ability to respond to cybercrime as something important."

5 Pilot Study

An initial version of the Cyber Resilient Behavior Questionnaire was refined based on experts' feedback and on validity testing of the survey on a small group of SMEs. The revised instrument has 42 statements, measured on a six-point Likert-type scale with no mid-point (ranging from strongly disagree to strongly agree). Even-numbered Likert scales force the respondent to commit to a certain position even if the respondent may not have a definite opinion [40]. Statements assess capacity, opportunity, and motivation regarding the performing of resilient functions to anticipate, monitor, respond, and learn within the organizational context. The statements focus predominantly on positive protective intentions of employees. This focus was chosen because respondents are probably willing to reveal these behaviors in a survey, yielding usable and reliable data (see also [19]). To reduce the chance of response bias, which is the tendency to favor one response over others [41], an - 'I do not know' - option was included for each statement. To further avoid response bias, both positively- and negatively-worded items were used in the survey [42]. Negatively-worded items may act as "cognitive speed bumps that require respondents to engage in more controlled, as opposed to automatic, cognitive processing" [43].

Table 1. Conceptual framework of resilient behavior. The left column shows the four generic resilience functions. The consecutive columns specify the abilities for resilient behavior for each of the three sources of behavior

	Capability	Opportunity	Motivation
Anticipate	Knowing what to expect	Having resources to look for developments further into the future	Willing to look for potential disruptions, novel demands or constraints, new opportunities, or changing operating conditions
Monitor	Knowing what to look for	Having resources to monitor the system's own performance as well as what happens in the environment	Willing to monitor that which is or could seriously affect the system's performance in the near term, positively or negatively
Respond	Knowing what to do	Having resources that help in taking prepared actions	Willing to respond to regular and irregular changes, disturbances, and opportunities
Learn	Knowing what has happened	Having resources to learn the right lessons from the right experience	Willing to learn from experience

Next, a pilot study was conducted, and the results were examined to identify any remaining problematic items and to establish the reliability of the main components of the survey. A total of 56 SME employees completed the pilot version of our instrument. All were high-level representatives at different SMEs. Cronbach's alpha was used as a measure of the internal consistency of the survey. This refers to the degree to which the items measure the same underlying construct, and a reliable scale should have a Cronbach's alpha coefficient above 0.70 [44]. Cronbach's alpha coefficients for each of the four main functions (i.e., Anticipate, Monitor, Respond, and Learn) all exceeded this recommended value. A series of Pearson product moment correlations were calculated to further assess the relationship between the items used to create the three main constructs. An examination of the correlation matrices revealed that all items significantly correlated at 0.3 or above with the other items in that construct.

Although the main focus of the pilot study was to test the instrument and the framework, the survey also included a set of questions concerning the incidence of cybercrime and victimization. Respondents were asked if their SME had been confronted over the last 12 months with cyberattacks and whether harms or costs were involved. We now present some preliminary results from our survey.

Almost half of the SMEs in our sample (48% or 22 SMEs) had been the victim of at least one cyberattack in the last 12 months. Phishing (32%) and viruses (11%) were reported most. Seven SMEs (12%) reported that damage was caused, for instance in the form of financial damage or business disruption. The overall score for cyber resilience of the SMEs was considered poor to average. SMEs scored 3.5 on the six-point Likert type scale. SMEs in our sample were best at responding to cyberattacks, and worst at learning from attacks (3.7 and 3.2, respectively). Because of the small sample size, no analyses were performed at the behavioral level for each of these functions. Nevertheless, this pilot study clearly shows that there is ample room for improvement in the cyber resiliency of the SMEs in our sample.

6 Discussion and Way Ahead

The cyber security field is in need of techniques to evaluate and compare the security design of organizations [8]. Many techniques have been proposed and explored, but these typically focus on auditing systems and technologies rather than on people. Our work is aimed at measuring cyber resilience of organizations through its employees rather than just with the technologies on which they rely [33]. Our framework gives an overview of relevant cyber resilient behaviors of employees within organizations. Accordingly, our conceptual framework allows for better development of human aspects of cyber resilience. It provides organizations with diagnostic capability to better prepare themselves for emerging cyber threats, while assuring the viability of those cyber assets critical to their business continuity [45]. Moreover, knowing what sources of behavior play a role in being cyber resilient, may help in the development of successful behavior change intervention programs. In future work, investigating how to link behavior change interventions to resilient behavior of employees, might prove important.

The Cyber Resilient Behavior Questionnaire is intended to investigate the capabilities, opportunities and motivation of people from all levels and functions of the organization. Many cyber security measurement tools used by organizations rely on information from only one or few organizational members, often specialists or managers responsible for cyber security [33]. This produces biased results, based on a single or few experiences, often with a vested interest in the results or scores achieved. The results that are produced with the Cyber Resilient Behavior Questionnaire are based on responses from a significant number of the company's employees. It is therefore more likely to tell us what the organization is actually doing and whether measures and policies have been embedded in the organization's social system [33]. However, in our pilot study, for practical reasons, only responses from high-level representatives from a small sample of SMEs were collected. Future research would benefit from the use of a larger sample of employees from all levels within organizations. Moreover, the SMEs in our sample were mostly small retailers. It is essential to validate our framework and instrument within other categories of the economy as well, for instance with large businesses in the industrial or financial sector.

Funding and Acknowledgments. This work was partially supported by the municipality of The Hague. The authors would like to thank Dr. Susanne van 't Hoff - de Goede, Michelle Ancher, Iris de Bruin and students from HBO ICT at THUAS for their assistance with this research effort. Further we would like to thank Dr. Jan Maarten Schraagen and Dr. Heather Young for their thoughtful and detailed comments that greatly improved the quality and readability of the manuscript. We are also grateful to the SMEs who agreed to participate in the surveys.

References

1. Ponemon Institute: Cost of cybercrime study (2017). https://www.accenture.com/t20171006T095146Z__w__/us-en/_acnmedia/PDF-62/Accenture-2017CostCybercrime-US-FINAL.pdf#zoom=50
2. Ponemon Institute: 2016 Cost of Cyber Crime Study & the Risk of Business Innovation (2016). https://www.ponemon.org/local/upload/file/2016%20HPE%20CCC%20GLOBAL%20REPORT%20FINAL%203.pdf
3. Accenture: Gaining ground on the cyber attacker. State of Cyber Resilience (2018). https://www.accenture.com/t20180416T134038Z__w__/us-en/_acnmedia/PDF-76/Accenture-2018-state-of-cyber-resilience.pdf#zoom=50
4. DiMase, D., Collier, Z.A., Heffner, K., Linkov, I.: Systems engineering framework for cyber physical security and resilience. Environ. Syst. Decis. **35**(2), 291–300 (2015)
5. Woods, D.D.: Four concepts for resilience and the implications for the future of resilience engineering. Reliab. Eng. Syst. Saf. **141**, 5–9 (2015)
6. Brown, C., Seville, E., Vargo, E.: Measuring the organizational resilience of critical infrastructure providers: a New Zealand case study. Int. J. Crit. Infrastruct. Prot. **18**, 37–49 (2017)
7. Parsons, K.M., Young, E., Butaviciu, M.A., Mc Cormac, A., Pattinson, M.R., Jerram, C.: The influence of organizational information security culture on information security decision making. J. Cogn. Eng. Decis. Mak. **9**(2), 117–129 (2015)
8. Bowen, P., Hash, J., Wilson, M.: Information Security Handbook: A Guide for Managers-Recommendations of the National Institute of Standards and Technology (2012)
9. Cain, A.A., Edwards, M.E., Still, J.D.: An exploratory study of cyber hygiene behaviors and knowledge. J. Inf. Secur. Appl. **42**, 36–45 (2018)
10. Yoon, C., Hwang, J.W., Kim, R.: Exploring factors that influence students' behaviours in information security. J. Inf. Syst. Educ. **23**(4), 407 (2012)
11. Leukfeldt, E.R.: Phishing for suitable targets in the Netherlands: routine activity theory and phishing victimization. Cyberpsychol. Behav. Soc. Netw. **17**(8), 551–555 (2014)
12. Leukfeldt, E.R., Kleemans, E.R., Stol, W.P.: A typology of cybercriminal networks: from low-tech all-rounders to high-tech specialists. Crime Law Soc. Change **67**(1), 21–37 (2017)
13. Young, H., van Vliet, T., van de Ven, J., Jol, S., Broekman, C.: Understanding human factors in cyber security as a dynamic system. In: International Conference on Applied Human Factors and Ergonomics, pp. 244–254. Springer, Cham (2018).
14. Bulgurcu, B., Cavusoglu, H., Benbasat, I.: Roles of information security awareness and perceived fairness in information security policy compliance. In: Proceedings of the AMCIS, pp. 419–430 (2009)
15. Dodge, R.C., Carver, C., Ferguson, A.J.: Phishing for user security awareness. Comput Secur. **26**(1), 73–80 (2007)
16. Talib, S., Clarke, N.L., Furnell, S.M.: An analysis of information security awareness within home and work environments. In: Proceedings of the International Conference on Availability, Reliability, and Security, pp. 196–203 (2010)
17. Crossler, R.E., Bélanger, F., Ormond, D.: The quest for complete security: an empirical analysis of users' multi-layered protection from security threats. Inf. Syst. Front. 1–15 (2017)
18. Da Veiga, A., Eloff, J.H.: A framework and assessment instrument for information security culture. Comput. Secur. **29**(2), 196–207 (2010)
19. Stanton, J.M., Stam, K.R., Mastrangelo, P., Jolton, J.: Analysis of end user security behaviors. Comput. Secur. **24**(2), 124–133 (2005)

20. Winnefeld Jr., J.A., Kirchhoff, C., Upton, D.M.: Cybersecurity's human factor: lessons from the Pentagon. Harv. Bus. Rev. **93**(9), 87–95 (2015)
21. Windle, G., Bennett, K.M., Noyes, J.: A methodological review of resilience measurement scales. Health Qual. Life Outcomes **9**(1), 8 (2011)
22. Linkov, I., Anklam, E., Collier, Z.A., DiMase, D., Renn, O.: Risk-based standards: integrating top–down and bottom–up approaches. Environ. Syst. Decis. **34**(1), 134–137 (2014)
23. Cox Jr., L.A.: Some limitations of "risk=threat x vulnerability x consequence" for risk analysis of terrorist attacks. Risk Anal. **28**, 1749–1761 (2008)
24. Frick, D.E.: The fallacy of quantifying risk. Def. AT&L **228**, 18–21 (2012)
25. Matzenberger, J.: A novel approach to exploring the concept of resilience and principal drivers in a learning environment. Multicultural Educ. Technol. J. **7**(2/3), 192–206 (2013)
26. Cutter, S.L., et al.: A place-based model for understanding community resilience to natural disasters. Glob. Environ. Change **18**(4), 598–606 (2008)
27. Hollnagel, E.: RAG – the resilience analysis grid. In: Hollnagel, E., Pariès, J., Woods, D.D., Wreathall, J. (eds.) Resilience Engineering in Practice. A Guidebook. Ashgate, Farnham (2011)
28. Van der Beek, D., Schraagen, J.M.: ADAPTER: analysing and developing adaptability and performance in teams to enhance resilience. Reliab. Eng. Syst. Saf. **141**, 33–44 (2015)
29. McDonald, N.: Organisational resilience and industrial risk. In: Hollnagel, E., Woods, D.D., Leveson, (eds.) Resilience Engineering, pp. 155–180. CRC Press, Boca Raton (2006)
30. Hollnagel, E.: Introduction to the Resilience Analysis Grid (RAG) (2015). http://erikhollnagel.com/onewebmedia/RAG%20Outline%20V2.pdf
31. Parsons, D.: National Organisational Resilience Framework Workshop: The Outcomes. National Organisational Resilience Framework Workshop (2007). http://www.tisn.gov.au/Documents/FINAL1Workshop.pdf. Accessed 22 Nov 2012
32. McManus, S., Seville, E., Vargo, J., Brunsdon, D.: Facilitated process for improving organizational resilience. Nat. Hazards Rev. **9**(2), 81–90 (2008)
33. Lee, A.V., Vargo, J., Seville, E.: Developing a tool to measure and compare organizations' resilience. Nat. Hazards Rev. **14**(1), 29–41 (2013)
34. Ferreira, P., Clarke, T., Wilson, J.R., et al.: Resilience in rail engineering work. In: Hollnagel, E., Paries, J., Woods, D.D., Wreathall, J. (eds.) Resilience in Practice, pp. 145–156. Ashgate, Aldershot (2011)
35. NIST: Framework for Improving Critical Infrastructure Cybersecurity, v 1.1, April 2018. https://nvlpubs.nist.gov/nistpubs/CSWP/NIST.CSWP.04162018.pdf
36. MacInnis, D.J., Moorman, C., Jaworski, B.J.: Enhancing and measuring consumers' motivation, opportunity, and ability to process brand information from ads. J. Mark. **55**, 32–53 (1991)
37. Michie, S., Van Stralen, M.M., West, R.: The behaviour change wheel: a new method for characterising and designing behaviour change interventions. Implement. Sci. **6**(1), 42 (2011)
38. Parsons, K., McCormac, A., Butavicius, M., Pattinson, M., Jerram, C.: Determining employee awareness using the human aspects of information security questionnaire (HAIS-Q). Comput. Secur. **42**, 165–176 (2014)
39. Parsons, K., Calic, D., Pattinson, M., Butavicius, M., McCormac, A., Zwaans, T.: The human aspects of information security questionnaire (HAIS-Q): two further validation studies. Comput. Secur. **66**, 40–51 (2017)
40. Brown, J.D.: What issues affect likert- scale questionnaire formats? JALT Test. Eval. SIG **4**, 27–30 (2000)

41. Randall, D.M., Fernandes, M.F.: The social desirability response bias in ethics research. J. Bus. Ethics **10**(11), 805–817 (1991)
42. Spector, P.E.: Summated Rating Scale Construction: An Introduction, no. 82. Sage, Thousand Oaks (1992)
43. Chen, Y.H., Rendina-Gobioff, G., Dedrick, R.F.: Detecting Effects of Positively and Negatively Worded Items on a Self-Concept Scale for Third and Sixth Grade Elementary Students (2007). Online Submission
44. Cronbach, L.J.: Coefficient alpha and the internal structure of tests. Psychometrika **16**(3), 297–334 (1951)
45. Linkov, I., Eisenberg, D.A., Plourde, K., Seager, T.P., Allen, J., Kott, A.: Resilience metrics for cyber systems. Environ. Syst. Decis. **33**(4), 471–476 (2013)

An International Extension of Sweeney's Data Privacy Research

Wayne Patterson[1(✉)] and Cynthia E. Winston-Proctor[2]

[1] Patterson and Associates,
201 Massachusetts Ave NE, Washington, DC 20002, USA
waynep97@gmail.com
[2] Department of Psychology, Howard University,
Washington, DC 20059, USA
cewinston@howard.edu

Abstract. About 20 years ago, the surprising research by Latanya Sweeney demonstrated that publicly available database information exposed the overwhelming percentage of United States residents to information easily available in order to facilitate the capture of such personal information, through techniques we now refer to as "dumpster diving." In particular, her research demonstrated that approximately 87% of the United States population can be identified uniquely using only the five-digit postal code, date of birth (including year), and gender. Although this result has held up over time, given the demographic parameters used in developing this estimate, Sweeney's technique made no attempt to develop similar estimates for other countries. In this paper, we use Sweeney's technique in order to provide estimates of the ability of similar demographics to provide the same type of data in a number of other countries, particularly those that tend to be as susceptible to data privacy attacks as the United States.

Keywords: Data privacy · International · Population · Life expectancy · Postal codes

1 Introduction

It is increasingly clear that two phenomena have grown extensively over the past two decades: first, the exponential growth of cyber-attacks in virtually every computing environment; and second, public awareness (whether accurate or not) of one's vulnerability to attacks that may be directly aimed at the individual, or more generally to an organization that maintains widespread data on the entire population.

The pioneering research of Dr. Latanya Sweeney demonstrated the vulnerability of most residents and computer users in the United States to the easily available demographic data necessary to identify sensitive information about any individual [1]:

"It was found that 87% (216 million of 248 million) of the population of the United States had reported characteristics that likely made them unique based only on {5-digit ZIP, gender, date of birth}." [1].

T. Ahram and W. Karwowski (Eds.): AHFE 2019, AISC 960, pp. 28–37, 2020.
https://doi.org/10.1007/978-3-030-20488-4_3

However conclusive was Sweeney's research concerning the citizens and residents of the United States, her research only provided a template for developing similar estimates regarding other countries throughout the world.

It is our purpose to extend the previous research to develop similar estimates regarding residents' vulnerability to data attacks using similar demographic data. In the first part, we will explore, by estimates of computer and Internet usage, and the prevalence of cyber-attacks, in many world countries.

Based on such estimates of usage and vulnerability, we will explore comparable demographic data for a number of target countries by utilizing population, life expectancy, and postal coding systems in the selected countries.

The value of the Sweeney research has been to introduce residents of the United States of the ease by which they can be identified in various databases and hence how their personal information and be captured via techniques known generally as "social engineering" or "dumpster diving." Since approximately 87% of the US population can be identified uniquely by only three pieces of (usually) easily found data, the identity of the individual can easily be compromised by persons seeking that information in publicly available databases.

In order to achieve the objectives of this paper, we will examine the feasibility of obtaining similar data or persons in a selection of other countries.

2 Selection of Countries for Analysis

We will begin by developing a methodology to select other countries for which this analysis can be performed. First, consider the level of concern in countries throughout the world in terms of the susceptibility to cyberattacks to discover personal information. Although the level of attacks is rising in virtually every country, we postulate that the level of concern by an individual citizen in a given country may be related to the widespread availability of national computer usage and Internet usage. These data will be demonstrated in Table 1 using data for the 30 countries with the greatest prevalence of computer availability and Internet usage, both in terms of the total numbers and the percentage of the population. It is part of our hypothesis that if a relatively small percentage of a country's population operate in cyberspace, there will be less interest either among the country's residents in terms of protecting their personal data; and by the same token, interest amongst those involved in cyber-attacks in finding personal data, since it might apply only to a very small percentage of the country's population.

The methodology for the selection of countries to be considered in this analysis is as follows: three statistics were identified for virtually all world countries. The first is each country's current population; the second, the percentage of the population owning or using computers; and third, the percentage of the population with access to the Internet. The use of these three statistics will generally give some measure as to the possibility of identity theft, whether done locally or via the Internet. For each of these three statistics, the countries of the world are ranked from the highest value to the lowest; then, if for country X, $X_{poprank}$ will represent the country's rank in world population, from highest to lowest among world countries [2]; $X_{internetuse}$, the percentage of Internet usage in the country, also from highest to lowest [3]; and finally

Table 1. Thirty leading countries in computer/internet use and cyberattack vulnerabililty.

Country	$X_{poprank}$	$X_{internetuse}$	$X_{computeruse}$	$X_{vulnerability}$
Japan	11	5	15	31
Great Britain	21	10	11	42
Germany	16	8	20	44
South Korea	27	14	14	55
United States	3	3	54	60
France	22	12	29	63
Russia	9	6	53	68
Canada	38	22	19	79
Spain	30	19	35	84
Brazil	5	4	90	99
Australia	51	32	23	106
Mexico	10	7	92	109
China	1	1	109	111
Netherlands	61	38	18	117
Malaysia	45	29	45	119
Argentina	31	23	70	124
Philippines	13	11	101	125
Poland	37	28	62	127
Italy	23	20	86	129
Turkey	19	15	95	129
Taiwan	54	35	41	130
Saudi Arabia	41	30	61	132
Sweden	75	50	16	141
Iran	18	17	107	142
Belgium	69	47	28	144
India	2	2	143	147
United Arab Emirates	80	52	17	149
Kazakhstan	58	41	51	150
Vietnam	15	16	120	151
Colombia	29	27	97	153

$X_{computeruse}$, representing the percentage of computer ownership or use, also ranked from highest to lowest. Thus for the residents of any country, $X_{vulnerability}$ is the sum

$$X_{vulnerability} = X_{poprank} + X_{internetuse} + X_{computeruse}.$$

In order to determine a rank order for countries being considered, define a development index based on the rank country among world countries using three orderings: first, the rank of world countries by population (2018 projections); second, the percentage of the population with Internet access; and third, the percentage of the population with local computer usage. To develop the overall ordering or countries that we wish to analyze, each of the three orderings above is added for each country, then all countries in our consideration are ranked by the lowest of the sums. In other words the optimal rank for a country would be if it was ranked first in each of three categories, and thus the sum of the three. In this analysis, the first country in combined ordering is Japan, with its three indicators adding to 31.

$$\text{Japan}_{\text{vulnerability}} = \text{Japan}_{\text{poprank}} + \text{Japan}_{\text{interneteuse}} + \text{Japan}_{\text{computeruse}}. = 11 + 5 + 15 = 31.$$

3 Prevalence of Cyber Attacks

In addition, there exists data to give an approximation of cyber-attacks on personal information, either through Internet attacks or through local attacks. We add to our selection of countries the 20 additional countries (as measured by Symantec [4]) with the greatest percentage of incidents of both online and local attack.

In a separate analysis, given data as collected by Symantec on the 20 countries with the highest percentage of local infections and the highest percentage of online infections, we performed a separate analysis based on these countries.

In order to calculate the analysis replicating the Sweeney study, one component not available in certain countries is a national postal code system which has been critical in the study regarding US demographics. Consequently, those countries which would fit our criteria otherwise analysis but have no national postal code system are listed in a separate table.

Table 2. Measures of malicious activity for 20 countries (from symantec)

Rank	Country	% of malicious activity:	Malicious code rank:	Spam zombies rank:	Phishing web hosts rank:	Bot rank:	Attack origin rank:
1	United States	23%	1	3	1	2	1
2	China	9%	2	4	6	1	2
3	Germany	6%	12	2	2	4	4
4	Great Britain	5%	4	10	5	9	3
5	Brazil	4%	16	1	16	5	9
6	Spain	4%	10	8	13	3	6
7	Italy	3%	11	6	14	6	8
8	France	3%	8	14	9	10	5
9	Turkey	3%	15	5	24	8	12
10	Poland	3%	23	9	8	7	17
11	India	3%	3	11	22	20	19

(*continued*)

Table 2. (*continued*)

12	Russia	2%	18	7	7	17	14
13	Canada	2%	5	40	3	14	10
14	South Korea	2%	21	19	4	15	7
15	Taiwan	2%	11	21	12	11	15
16	Japan	2%	7	29	11	22	11
17	Mexico	2%	6	18	31	21	16
18	Argentina	1%	44	12	20	12	18
19	Australia	1%	14	37	17	27	13
20	Israel	1%	40	16	15	16	22

Using the data from the original 30 countries selected, as well as the additional 22 from the Symantec data, we will attempt to perform a simplified analysis, comparable to the work by Sweeney, in order to develop national comparisons of the percentage or numbers of persons in the population of each country in terms of the uniqueness of identification, using factors similar to Sweeney's: gender, date of birth (including year), and postal code of residence.

It should be noted that in a number of the countries that we select in terms of the criteria described above, there is no postal code system, and thus a comparable analysis cannot be performed. We note that for our further analysis, we choose (for space reasons) the leading 20 countries identified in Table 1 in terms of attack vulnerability. In addition to these 20, we note that 18 also appear in Table 2, for the actual susceptibility to attack. However, it is noted that all of the leading 20 countries in Table 1, the only countries in Table 2 but not in Table 1 are India and Israel.

4 Postal Code Systems

To replicate the Sweeney study for other countries, it is necessary to identify the total population, the life expectancy by country, and the postal code system in such countries.

The first two are easily found and have a high degree of accuracy. The existence of the postal code system, which does exist in most countries but not all; is of a different nature, since the information that is easily available is the potential range of values for postal codes in all of our selected countries. For example, and using 'N' to represent decimal digits in a postal code, and 'A' for alphabetical characters, it is possible to determine the total range of possible postal codes. For example, in the United States five-digit ZIP code system, which we would indicate as "NNNNN", there are a total of 10^5 = 100,000 possible postal codes. However, as reported by Sweeney at the time of her research, only 29,000 of the possible five-digit combinations were actually in use. (The corresponding use of US ZIP code numbers at present is 40,933.)

Most of these data have been compiled for approximately 200 countries, but in order to apply the Sweeney criteria, we limit the further analysis to a smaller set of countries.

In order to develop a comparison in terms of the privacy considerations in individual countries, is necessary to be able to estimate the key statistics Sweeney used. Population data is easily available for all United Nations member countries, as are mortality rates or life expectancy rates to develop applicable birthdays as in Sweeney's paper. However, the third statistic used by Sweeney is not as widely available. This statistic is, for the United States, the 5-digit form of postal code, called in the US the "ZIP code". It is noted that in the US context, that most if not all postal service users have a 9-digit ZIP Code, sometimes called the "ZIP+4", NNNNN-NNNN, but the recording and storage of US postal codes still varies widely, and most databases that might be discovered by a hacker would only have the 5-digit version, "NNNNN". Furthermore, Sweeney's original research only used the original 5-digit ZIP Code.

In our comparative study, we have had to eliminate a number of United Nations countries that either do not have the postal code system, or it is not readily available. This seems to be only in a relatively small number of United Nations countries.

The other complicating factor in this comparative study is that in most countries, there is a distinction between the characters of potential postal codes as a function of the syntax of the structure of postal code assignment. Throughout the world, most postal codes use a combination of numerals {0, 1, 2, …, 9} which we describe as 'N'; and letters of the relevant alphabet. In the Roman alphabet (mostly in uppercase), we have {A, B, C, …, Z} which we designate as 'A'.

In the case of the older US 5-digit ZIP Code, the syntax is NNNNN, which allows for the maximum possible number of ZIP Codes as $10^5 = 100{,}000$. As a comparison, the Canadian postal code system is ANA NAN, therefore $26^3 \times 10^3 = 17{,}576{,}000$.

Thus our level of analysis in estimating the actual number of postal codes is simply to use the calculated level based on the syntactical postal codes. To be more accurate, however, it is necessary to take into account that many postal systems restrict the usage of some of these symbols for perhaps local reasons. Thus, to obtain a more precise comparison, it is important for possible to determine the actual number of postal code values actually in use, as opposed to the number theoretically in use.

For example, the current estimate of US ZIP Code numbers in use is 40,933, or 41% of the allowable values. These estimates are only available for a smaller number of countries.

In order to determine a "Sweeney Index" for our 32 selected countries (30 from Table 1, India and Israel from Table 2), we must first determine the life expectancy by country, and the number of possible values in the country's postal code system.

The first challenge in this analysis arises because not all countries have a postal code system. The following table demonstrates the status of postal codes in our 30 selected countries. In most cases where a postal code system exists, it is defined by a numerical sequence, ranging from four, five, six or even nine digits; and often also by several alphabetic symbols, the most part using the Roman alphabet. In the table below the use of a numeric character is indicated by N, and an alphabetic character by A. Thus, for example, a country using a five-digit number for the postal code would be represented in our table as "NNNNN".

The first estimate of the number of postal codes for country is determined by the syntax and the potential number of occurrences for each character in the string representing the code. In a number of cases, it is possible to determine if a country uses all of the possibilities for codes under its coding system. But in most countries, not all possibilities are used—only a certain fraction of the eligible set of codes are actually in use; unfortunately this information is not readily available for all countries.

The major conclusions by Sweeney are obtained by the analysis of internal United States data on ZIP Codes to approximate the distribution of active addresses as distributed over the entire set of postal code values. Sweeney defines several methods of distribution, including uniform distribution, which would certainly simplify calculations for other countries. It is likely to be less realistic than many other options; nevertheless, the scope of this article, we will only calculate the likelihood of unique identification of individuals assuming uniform distribution of individuals in countries; since we do not have access to the necessary internal postal code distributions in other countries. Nevertheless, we feel uniform distribution gives a reasonable first approximation to the Sweeney results.

Using the uniform distribution process, we can calculate the total number of "pigeonholes" accurately for many countries, and then the uniform distribution by dividing the population by the number of pigeonholes.

5 "Pigeonholes"

The problem then becomes the conducting of an assessment of the data for the number of persons that can fit into each of the potential categories, or "pigeonholes" in a frequently-used term in computer science. Another way of phrasing the conclusions of Sweeney's earlier study is to say that of all the pigeonholes, approximately 87% have no more than one datum (that is, no more than one person) assigned to that pigeonhole.

Table 3. Potential actual number of postal codes for certain countries.

Country	Postal code format	Maximal possible postal codes	Actual postal codes (p_p)	National population
Japan	NNNNNNNN	100,000,000	64,586	127,484,450
Great Britain	A A N AN	45,697,600	1,700,000	66,181,585
Germany	NNNNN	100,000	8,313	82,114,224
South Korea	NNNNN	100,000	63,000	50,982,212
United States	NNNNN	100,000	40,933	324,459,463
France	NNNNN	100,000	20,413	64,979,548
Russia	NNNNNN	1,000,000	43,538	143,989,754
Canada	ANA NAN	57,600,000	834,000	36,624,199

(continued)

Table 3. *(continued)*

Country	Postal code format	Maximal possible postal codes	Actual postal codes (p_p)	National population
Spain	NNNNN	100,000	56,542	46,354,321
Brazil	NNNNNNNN	100,000,000	5,525	209,288,278
Australia	NNNN	10,000	2,872	24,450,561
Mexico	NNNNN	100,000	100,000	129,163,276
China	NNNNNN	1,000,000	860,000	1,409,517,397
Netherlands	NNNN AA	250,000	5,314	17,035,938
Malaysia	NNNNN	100,000	2,757	31,624,264
Argentina	ANNNAAA	240,000	1,237	44,271,041
Philippines	NNNN	10,000	10,000	104,918,090
Poland	NN-NNN	100,000	21,965	38,170,712
Italy	NNNNN	100,000	4,599	59,359,900
Turkey	NNNNN	100,000	3,314	80,745,020
India	NNNNNN	1,000,000	153,343	1,339,180,127
Israel	NNNNNNN	10,000,000	3,191	8,321,570

The number of pigeonholes in Sweeney's study for the United States is calculated by the product of the potential number of persons identified by birthdate including year, gender, and 5-digit ZIP code. The contribution to this number related to gender is 2, say $p_g = 2$. For birthdate, we approximate the number of values using 365 for days of the year (a slight simplification ignoring leap years), multiplied by the number of years, estimated by the country's life expectancy in years [5]. Call this p_b. The final relevant factor in estimating the number of pigeonholes is the number of potential postal codes, p_p. Then the total number of pigeonholes is

$$\#\text{pigeonholes} = p_g \times p_b \times p_p = 2 \times (365 \times \text{life expectancy}) \times p_p$$

One remaining problem is the calculation of the number of postal codes, p_p [6]. It is an easy calculation to find the maximal value for p_p say p_{pmax}. For example, for the 5-digit US ZIP code system, that maximal value is $p_{pmax} = 10^5 = 100000$. At the time of Sweeney's research, the number of ZIP codes actually used was $p_p = 29343$ ([1], page 15), or 29.3% of the total number of ZIP code values. At present, the number of ZIP codes in use is 40,933.

Given available data for all world countries, the value p_p is often not made public (Table 3).

6 Comparing the Privacy Issues

It is not possible with most countries perform the same detailed analysis as can be found in the work of Sweeney, since for many countries the detailed data she obtained is not available. However, several other analyses are possible which can still provide considerable insight with the comparable Sweeney analyses and hence the state of privacy issues in certain countries.

The two analyses we have performed use the following:

1. Assumption of uniform distribution of Sweeney parameters with postal codes using theoretical postal code values.
2. Assumption of uniform distribution of Sweeney parameters with postal codes using actual postal code values.

We have been able to perform the analyses for most United Nations countries where appropriate postal information is available. However, for the purposes of this paper, we have restricted to those 30 countries we identified in Sect. 1 with the potential of the greatest risk because of greatest computer and Internet penetration.

7 Conclusions

The statistic in the final column of Table 4 provides the key information for our conclusions. The meaning of the statistic "Average Number of Persons/Pigeonhole", which we will abbreviate AvgPP, is an indication of the potential lack of individual privacy for computer users in the country in question. We note first that the calculation for the US, 0.1369, is the level by which a single individual may not be identified; thus 1 − AvgPP = 0.8631 is very close to the statistic computed by Sweeney in her 2001 paper, that 87% of the US population could be uniquely identified by the Sweeney criteria; in other words, only 13% of the US population could not be so identified. Rather than a direct confirmation, this does demonstrate a consistency even with an 18-year difference in the date and a change in the US population (248,418,140 in 2000 and 324,459,463 at present) and in the number of used ZIP codes (29,343 in 2000 and 40,933 at present).

Table 4. Average number of residents per Sweeney criterion (pigeonholes) for 22 countries

Country	Gender x days/yr	Life expectancy	Actual postal codes	Pigeonholes	Population	Average no. of persons/pigeonhole
Great Britain	730	81.2	1,700,000	1.008E+11	6.62E+07	0.0007
Canada	730	82.2	834,000	5.005E+10	3.66E+07	0.0007
South Korea	730	82.3	63,000	3.785E+09	5.10E+07	0.0135
Spain	730	82.8	56,542	3.418E+09	4.64E+07	0.0136
Mexico	730	76.7	100,000	5.599E+09	1.29E+08	0.0231
China	730	76.1	860,000	4.778E+10	1.41E+09	0.0295
Poland	730	77.5	21,965	1.243E+09	3.82E+07	0.0307

(continued)

Table 4. (*continued*)

Country	Gender x days/yr	Life expectancy	Actual postal codes	Pigeonholes	Population	Average no. of persons/pigeonhole
Japan	730	83.7	64,586	3.946E+09	1.27E+08	0.0323
Israel	730	82.5	3,191	1.922E+08	8.32E+06	0.0433
France	730	82.4	20,413	1.228E+09	6.50E+07	0.0529
Netherlands	730	81.9	5,314	3.177E+08	1.70E+07	0.0536
Russia	730	70.5	43,538	2.241E+09	1.44E+08	0.0643
United States	730	79.3	40,933	2.370E+09	3.24E+08	0.1369
Australia	730	82.8	2,872	1.736E+08	2.45E+07	0.1408
Germany	730	81.0	8,313	4.915E+08	8.21E+07	0.1671
India	730	68.3	153,343	7.646E+09	1.34E+09	0.1752
Malaysia	730	75.0	2,757	1.509E+08	3.16E+07	0.2095
Philippines	730	68.5	10,000	5.001E+08	1.05E+08	0.2098
Italy	730	82.7	4,599	2.776E+08	5.94E+07	0.2138
Turkey	730	75.8	3,314	1.834E+08	8.07E+07	0.4403
Argentina	730	76.3	1,237	6.890E+07	4.43E+07	0.6425
Brazil	730	75.0	5,525	3.025E+08	2.09E+08	0.6919

The meaning of the ordering in the AvgPP Column is to show that in which countries personal privacy is even less than in the United States (again 87% identifiable by the three Sweeney criteria). These are the 12 countries above the United States in the Table 4, with the Great Britain having the least down to Israel; and the other 9 below the United States having a greater level of personal privacy, from Australia to Brazil.

This paper has implications for future behavioral cyber security problem specification, theoretical conceptualization, and methodological development.

References

1. Sweeney, L: Simple Demographics Often Identify People Uniquely. Carnegie Mellon University, Data Privacy Working Paper 3, Pittsburgh (2000)
2. United Nations. https://population.un.org/wpp/Download/Standard/Population/
3. Wikipedia. https://en.wikipedia.org/wiki/List_of_countries_by_number_of_Internet_users9)
4. Symantec, Internet Security Threat Report. ISTR 2018 vol. 3. http://resource.elq.symantec.com
5. World Health Organization. http://apps.who.int/gho/data/node.main.688?lang=en
6. Wikipedia. https://en.wikipedia.org/wiki/List_of_postal_codes

The Human Factor in Managing the Security of Information

Malgorzata Wisniewska, Zbigniew Wisniewski(✉), Katarzyna Szaniawska, and Michal Lehmann

Faculty of Management and Production Engineering, Lodz University of Technology, ul. Wolczanska 215, 90-924 Lodz, Poland
{malgorzata.wisniewska, zbigniew.wisniewski}@p.lodz.pl, szaniawska.katarzyna.m@gmail.com, michal.lehmann@pekao.com.pl

Abstract. The article discusses the findings of analyses of information security levels in Polish universities and presents the results of comparative analyses performed currently and those of security level studies conducted in 2008. The re-examination has demonstrated that there has been no significant improvement in the level of security. Despite the increase in public awareness of threats on the Internet and the increase in general public's ability to protect their own information resources, the analogous trend in universities has not occurred. Resources are still not adequately protected. The authors present the results of the comparative analysis and try to explain this paradox in conclusion.

Keywords: Information · Security management · Information security threats · Human factor · Organizational culture in information security management

1 Introduction

In terms of information security, most organizations target external threats, such as viruses, even though experiential data indicates that threats originating from within companies are more probable and more dangerous.

The fact that there is no public debate on internal malpractice is not synonymous with the fact that it does not exist. Cases of such incidents are simply not disclosed or, worse still, remain unnoticed.

Among the major threats to information security in organizations, the ones which receive most attention are viruses, Trojan horses, and "Internet worms", whereas it is only as secondary threat that the improper conduct of employees is mentioned [2]. It remains problematic that few organizations recognize the fact that relevant training and raising employees' awareness of information security are among the key success factors in the area of asset security.

Malpractice statistics show that external attacks are a relatively rare phenomenon, while the majority of security breaches involve an organization's employees. There are a number of underlying causes, ranging from poor security training skills, low staff motivation, disastrous work environment, to the fact that employees simply must be

T. Ahram and W. Karwowski (Eds.): AHFE 2019, AISC 960, pp. 38–47, 2020.
https://doi.org/10.1007/978-3-030-20488-4_4

granted access to information required for the performance of their duties, and this can also result in malpractice (mainly due to recklessness).

The most serious breaches invariably exploit the human factor, which means that special emphasis must be placed on educating workers, and that security training programs must also include all manipulative techniques, the so-called social engineering techniques. In large companies, too much stress is placed on the sophisticated security of information systems, while a wide range of aspects that depend solely on individuals' good will and common sense is neglected.

There is evident a somewhat perplexing paradox in all of this - the authorities of the organization predominantly believe that they are more susceptible to attacks from without and not to the breaches from within the organization. Such a claim is made despite the fact that there have been made repeated and confirmed observations to the effect that threats caused by authorised network users – permanent personnel as well as temporary staff or former employees having access to resources located within the companies - are more likely and more dangerous in their repercussions.

A question needs to be raised as to how real this threat is. There is a lack of official data on losses caused by insiders' unauthorized actions, but abuses affect practically every company, regardless of geographical location, industry, size and they typically remain undetected [7]. According to surveys conducted among employees in the course of research on, among others, malpractice, one in five employees declares that they are aware of other employees robbing the employer. As the enterprise grows, the possibility of abuse increases while the detection rate drops [5].

The aim of this article is to present the levels of awareness of information security risks in such specific organizations as higher education institutions. The initial research was carried out in 2008, followed by a comparative study performed in the same entities in 2018 with a view to verifying whether any changes have occurred. The data from the latest survey will be presented and those from 2008 will be referred to for comparison. The research focused on public universities in Poland. The study was part of the research project of the SYDYN group.

2 Awareness of Threats

Consideration should be given to which risks are perceived by organizations as the most serious and how sensible and justified are the measures taken to avoid them?

In order to be able to secure information effectively, a company needs to understand what it wants to protect and how to protect this resource in the most vulnerable areas. This is the first step in the direction of initiating appropriate actions to effectively utilize the available resources.

Studies indicate that some enterprises are strengthening their existing control mechanisms in particularly sensitive areas [1]. Simultaneously, however, there are many symptoms of inadequate management of resources by, for example, assigning to them ineffective control mechanisms and neglecting the necessary actions in the most vulnerable aspects. The list of the most serious risks includes:

- malicious software (48% vs 58% w 2008) and
- unsolicited mass mailing (spam) – 65% (58%-2008).

These are the areas where the majority (83% of universities vs. 67% in 2008) claim that they have adequate protection in the form of antivirus systems or solutions protecting their networks against spam. Meanwhile, however, universities weaken the effectiveness of these security measures, neglecting activities aimed at raising employees' awareness, as to, for instance, what are the consequences if they open suspicious e-mail attachments [4].

The reason for this is simple to establish. The management, influenced by the media and suppliers, focuses their attention on the state-of-the-art in technology, which according to highly touted theories (with adequate support in the form of reviews and tests) are the only way to protect companies from leaks of information and viruses and so-called computer "worms" [3] with their media attention potential.

However, the price for this type of purchase is, or may become, completely disproportionate to the promised benefits. When buying expensive and widely advertised software, which is supposed to protect the company against loss of information, intrusions into systems, etc., company authorities are guided by the principle that "it is better to do anything than to do nothing". However, they often fail to notice the terrifying reality that it is not attacks from the outside, but those from the inside that pose the most serious threat. The most common reason for such short-sightedness is either lack of sufficient knowledge about the mechanisms controlling the information flow processes, or nonchalance manifested in ignoring signals about threats and methods of avoiding them, both of which can simply equaled to "lack of self-preservation instinct".

The illusory steps taken with regard to information security and the elimination of threats are also an indirect result of the fact that university authorities are reluctant to accept that they have fallen victim to an internal attack, which is evidently due to a false sense of security rather than to the low scale of the threat.

Unfortunately, in a precarious economic situation, the chance for an employee to obtain personal benefits offers additional motivation to commit abuse, which renders the threat even more real.

Persistent forms of intrusion, which have nothing to do with technical solutions, are rather common and unfortunately based mainly on human behaviour. The prevalence of security breaches from within stems from the fact that they do not require the use of complicated methods and usually occur during regular working hours. In some cases, dissatisfied employees simply want to damage the organization and its reputation. A common scenario is that access rights are not immediately revoked, so that former employees - including contractors and temporary workers - continue to enjoy access to network resources. Such a situation persists on average for more than two weeks after the employee's contract termination [4].

Despite the use of security mechanisms, numerous universities (35% vs. 38% in 2008) experienced incidents related to hardware or software failure, which resulted in a perceptible lack of availability of key IT systems. It is in such situations that one can speak of a double loss - firstly: the costs incurred for expensive security solutions did not bring the desired results, and secondly: no investment was made in a truly significant threat - the risk of internal attacks which can materialize at any time.

Safety is inextricably linked to costs, and universities are aware of the need to pay more for higher quality products. They require IT solution providers to run more and more rigorous tests to detect the vulnerability of their products before releasing them on the market, rather than utilize end users as beta-testers who will find errors on their own.

3 Human Factor in Information Security Management

Given that information is the most valuable commodity of our time, its theft has virtually become an everyday practice. The victims of acts of information theft, are, first of all, naivety and ignorance of the employees in an organization while various techniques of manipulating personnel are the most frequently deployed weapon. The application of trickery in order to extort business information from authorized employees is one of the means to obtain an organization's resources. Others employ more advanced techniques to hack information systems, and others, still, lead to intrusions into buildings and premises of the organization. Aside from the typical threats, both external and internal, whose causes can be attributed to failures, disruptions, and errors in information systems, the primary source of threats is the human factor. It is individuals who have the biggest impact on the system. Through their actions they can cause both hardware and software malfunctions, breach data sets and access other system resources. The type and magnitude of a potential threat depends on the function that a given person performs in the organization, his or her skills and, of course, whether he or she acts unintentionally or deliberately. Therefore, threats caused by human activities can be divided into unintentional and deliberate. Conducted research has demonstrated that the vast majority of unintentional threats (similarly as in 2008) are completely random and are usually the result of inattention, submission to manipulation, thoughtlessness, negligence or incompetence of individuals in the design or operation of the system.

Methods of manipulating personnel include, among others, sending falsified e-mails and impersonating other users whom the individual trusts by entering the person's address in the identification field and then including a request or order to send information or open attachments which, when activated, may initiate a process enabling remote access to the user's workstation (browsing catalogue contents, changing and deleting information, intercepting passwords or encryption keys). Unintentional threats are numerous and hardly predictable. Unfortunately, these threats cannot be eliminated by applying technical solutions. These problems have to be solved procedurally. The risk of their occurrence can be reduced by appropriate selection of staff, reasonable allocation of access rights, trainings, software and hardware security measures.

Quite different is the case of intentional and deliberate threats. A very significant hazard is internal attacks typically related to improper use of the organization's ICT system by users. Whereas the share of internal attacks in all recorded events is very high, the only effective protection against such events is a well-designed security policy.

Only a handful of universities (10%) have begun work on developing a training and awareness-building programs - a fundamental element in establishing an effective information security strategy. There is no doubt that it is difficult for employees to react to abuses committed in the organization if they have not been adequately trained in recognising such incidents and reporting their occurrence to the right personnel.

In most cases, an organization's management confirms that the users' lack of risk awareness is the main obstacle to achieving the required level of security.

Investment in technical solutions cannot completely eliminate the human risk factor. Even if senior management claims that information security is important, time, attention, actions taken and resources allocated to improve security are not adequate.

A number of observations of participants and interviews with representatives of universities were conducted. They revealed several categories of irregularities typical for the majority of the surveyed universities. In many cases, the respondents revealed that they suspect, rather than know, that in some aspects of information security policy certain steps are being implemented to the benefit of their units.

The results of research conducted by the authors for nearly 15 years indicate the following issues:

- management is reluctant to attach adequate importance to human capital and is much more willing to invest in technical solutions,
- fewer than half of the study participants provide systematic safety and control training to their staff.

The most frequent irregularities in the area of information protection are identical to those reported in 2008. It should be noted that despite considerable technical progress in the field of information security and the declared increase in awareness among system users and management, there has followed no significant improvement in security. Among the most common sources of risk, the following have been recognized.

1. **Incorrectly positioned screens at customer service workstations** – computer screens are positioned in such a way that unauthorized persons are able to read the displayed information (e.g. in the customer queue, persons can see the data of the customer being served and photograph them).
2. **Employees do not lock workstations with password when leaving the workstation** – this gives unauthorized persons easy access to information contained in the systems (especially in customer service rooms).
3. **Employees do not lock (do not physically secure) rooms when leaving** – there is easy access to information contained both in IT systems (with unsecured workstations) and any paper documents located in the room. Documents with confidential information about the organization as well as information concerning the organization's stakeholders may be stolen.
4. **Documents are stored in generally accessible rooms** - cabinets with documents (frequently unsecured) stand in an area accessible to everyone entering the room. Sometimes such cabinets stand outside a room, e.g. in a corridor. In customer service rooms these are frequently not cabinets per se, but shelves lined with documents.
5. **5. Internal attacks - involving the use of e.g. social engineering.** There is lack of awareness among employees that no information to which they have access should be given or disclosed to unauthorised persons. Employees, in particular departments that process information that is not accessible to everyone (e.g. financial,

human resources, IT, science, teaching, etc.), do not check the identity and authorization of those who ask for information.

6. **Movement around the organization's premises** - employees who have been made redundant and have ceased cooperation with the organization should not have access to office space. A barrier in the form of security is not effective if there is no proper flow of information within the organization. Security personnel and other employees are not informed that specific individuals no longer work for the organization and have no right of access to certain areas.
7. **Disclosure of information concerning employees of the organization** - this mainly involves posting internal information concerning the organization and employees in publicly accessible areas.
8. Authorization **regarding access to information systems** - lack of procedures for organizing matters related to granting, modifying and removing of authorization to information systems.
   a. **Employees are granted or modified authorization to the systems without written confirmation.** There is no evidence that granting privileges has been authorized. Unfortunately, such situations usually involve the most senior officials in organizations - they usually have urgent needs and believe that no regulations apply to them.
   b. **Clearances** in IT systems and network resources **are not withdrawn in due time** - e.g. upon termination of cooperation.
   c. Lack of uniform rules on the storage of information on clearance levels - **lack of a uniform system of authentication and authorization.**
   d. **The fourth case - authorizations for third parties.** It is frequent that outsourced personnel, service technicians, contract workers, etc. perform work for the organization. However, this area is unregulated, because e.g. the systems themselves do not offer an option of granting rights in a manner consistent with the internal procedures. In the majority of cases, there exist the so called special accounts with wide range of access rights which are used by service technicians, administrators, programmers, etc. This cannot be allowed to happen. Accounts must be individually created for everyone who is supposed to have access to the system - with a separate identifier. Otherwise, in the event of an incident it is impossible to know who did what and when.
9. **Sharing personal logins** – this is also a very frequent situation: an employee discloses his or her own system login and password to a colleague at work.
10. **Storing system passwords** - it is still true that employees store passwords to systems on post-it notes under the keyboard, desk drawers are full of slips of paper with specified systems, logins and passwords to them. There were reports of notebooks with entire password history in these drawers.
11. **Open wireless networks** - in most units (70% vs. 85% in 2008) there were examples of open wireless networks operating without any security measures. In many cases the coverage of these networks extended even beyond the university campus gounds.

4 The Role of Management in Information Security Administration

Organizations which understand that information protection must be an integral part of their business recognize the need for a proper application of a combination of different types of security measures and know that it is only when all control mechanisms are complementary that they can achieve the most desirable outcome. By using a well-designed security policy, the organization sets out clear rules for users. The policy defines the areas of permitted and prohibited activities. It appears valid to say that applying only organizational control mechanisms is, from an economic point of view, both cost-effective and simple to implement. Simpler, in any case, than technical safeguards. However, despite their apparent simplicity in terms of implementation steps, they constitute an immensely serious and complex element of the security structure and prove challenging when attempts are made to enforce them in the actual operating conditions of an organization. This is a major problem, especially if middle management support is lacking. A prolonged lack of consistency in the management's actions in this respect may lead to very costly consequences for the organization.

Senior management support is essential to ensure safety. However, mid-level managemers play a key role in the allocation of tasks to workers and the definition of their entitlements and responsibilities. Their absence in the process may result in the safeguards being either circumvented or simply disregarded. Rather few organizations can afford such a high risk.

Without the proper approach on the part of the management, one which is coherent in terms of responsibility, awareness, and education, no policy and contingency plans will guarantee the safety of an organization. A simple conclusion resulting directly from these considerations is that the role of the management in the organization is to create such awareness and such an organizational culture that every employee will feel that he or she is not only an insignificant cog in a huge machine, but that he or she has an impact on the machine functioning, development and safety. The employee must be aware of the tasks he or she performs at the assigned post and that even the slightest activity they perform has an impact on the functioning of the whole organization. Knowledge, which is frequently restricted to some individuals, should become available to the extent practicable in given circumstances with a view to limiting the willingness to verify it in an unauthorized and dangerous manner from the standpoint of information security.

Enforcement of the company's security policy means that everyone is responsible, and not only the persons appointed to the area of information security administration within the organization. It should also be remembered that successful implementation of a well-designed - and, what is more, effective - security policy can only be ensured if the security regulations are formulated in a clear manner. They should be so clear and unambiguous that their application does not pose any difficulty and that the actions related to ensuring the security of the information processed become obvious and instinctive [6].

Organizations which are seriously concerned about the effectiveness of control mechanisms know that they cannot merely address external threats. Although the public is easily excited about viruses and incidents involving external intrusions, it is the employees and authorized network users who can cause the most damage.

The increased emphasis on the application of the company leadership's security management model, which includes raising employee security awareness, should also entail a more thorough scrutiny of employees and business partners.

There is some irony to it. Human resources represent the greatest potential risk, but when provided with appropriate support, they can become the strongest element of protection within an organization. Assigning information security the right priority "at the top" is a key element of success. The right approach and motivation at the highest level of management can inspire radical changes. It is widely accepted that the perception of information security as a necessary cost to business appears obvious. However, looking at information security as a business growth factor, as an element of competitive advantage, and as a method of protecting the value of resources, requires a completely new and broader perspective. Senior management's perception of information security must therefore undergo a radical change. It is for university managers to create a culture of security based on each unit's awareness and sense of responsibility for their own actions. Creating such an organizational culture is challenging and can be achieved if senior management itself follows the rules it has established. If organizations do not make information security a part of their culture, it is likely that regulators will become involved in this process in the near future. With the increasing complexity of the business world, information security requirements are also increasing.

5 Conclusion

Managing security is an immensely important topic, yet it is frequently not addressed proportionally to the gravity of the issue. Regrettably, for university management, information security is very often synonymous with the security of information systems. It is true that at present it is difficult to find an organization in which information processing would not be supported by IT systems. This is probably the root cause of the tendency to reduce the issue of information security to the level of IT systems and concurrently to assign responsibility for information security to IT services in the organization. This is a very serious mistake and for this reason the approach of IT department employees is particularly striking, as on various occasions they reveal that they understand information security management mainly as security systems management. And yet, the security of information systems is only a part of the concept of information security (a rather narrow subset of it). There are still paper records in universities, despite the fact that IT systems have taken over most of the information processing operations. Documents are collected in cabinets and archives. Additionally, it should be borne in mind that information is also notes, diagrams and handwritten sketches, which are usually found in drawers and employees' desks. The multitude of these documents and the many types of media on which they are collected generates a serious problem - how to manage and control them? The answer is not difficult, one simply has to create a basis for effective information security management in the organization, similarly to the management of other processes necessary for the organization to function. The most effective tool is the Information Security Policy, which may also include many other policies [8]. One such policy is, of course, the IT Systems Security Policy. In addition to the management's declaration of what it wants to protect and for what reasons, this document also contains procedures, rules and instructions, i.e. documents which should state how to fulfil the declarations.

Equally important as the implementation of appropriate technological solutions, the creation and implementation of required procedures, such as training and raising employees' awareness of their responsibility for the security of processed information. The starting point for effective information security management in the company is the understanding that it does not consist only in the purchase and implementation of such or other security systems.

The indispensability related to the implementation of the Information Security Policy has its source in legal and business aspects, however, it is associated with the need to protect the organization against various internal and external threats. The fact that the threats are an inherent element of every area of the organization's operation should prompt the university authorities to reflect on an enormously important issue related to their identification and identification of their sources. Most often, however, it is so that the organization's activities related to conducting a thorough threat analysis are not the starting point for the creation of a security system. As a consequence, the developed protection system is imperfect, not to say flawed, and, in any case, may prove to be ineffective. This is due to the fact that if the most important weaknesses and threats are not adequately known, the only outcome of the activities related to the creation of a security system is to take into account the budget and qualifications of the system's developers, while neglecting the actual needs that result from the threats and, as a result, affect the final form and effectiveness of the designed security system.

The IT risk is growing at an alarming rate as the authorization levels and the number of users having access to information increase. Individuals operating within an organization believe that the risk of being caught for fraud is low, as they have detailed knowledge of the system, possible means of access and know organizational control mechanisms. They usually work for personal gain, but some can also launch completely selfless attacks. The recklessness or mere ignorance of others, weakening even the best controls and technical security mechanisms, may mean that such individuals may become the greatest threat to information security within an organization. Unfortunately, business managers too often underestimate this threat. They only mention it as the second most important.

The lack of progress in information security in universities compared to 2008 levels may be surprising. After all, more and more members of the general public use advanced IT tools on their own and know how important it is to protect their identity, image and financial assets, among others. Individuals know how to protect themselves and they do it better and better. Why is there no significant progress in information protection at universities? The probable reason is that in systems that we use privately and for our own purposes, the technological advancement is so great that system providers compete with each other very strongly as they are exposed to constant attacks from the press and Internet users on social networking sites. Therefore, they try to implement security mechanisms that are resistant to not only attacks, but also even to accidental loss of security. Such actions have made us, as users of IT systems, less sensitive to threats. And this is paradoxical. It is for this reason that the university staff perform risky operations and do not make sure appropriate preventive mechanisms are in place. They simply use private life habits in the organization, i.e. they are not sensitive enough to threats.

References

1. Andersen, R.: Inzynieria zabezpieczen. WNT, Warszawa (2005)
2. Lukatsky, A.: Wykrywanie wlaman i aktywna ochrona danych. Helion, Warszawa (2004)
3. Pakula, J.: Bezpieczny komputer. Czyli jak chronic sie przed hakerami, wirusami, spywarem, spamami itd., Help (2005)
4. Pieprzyk, J., Hardjono, T., Seberry, J.: Teoria bezpieczenstwa systemow komputerowych. Helion (2005)
5. Raport Ernst & Young: "Swiatowe badanie bezpieczenstwa informacji 2015"
6. Wisniewska, M.: Influence of organizational culture of a company on information security, In: Lewandowski, J., Kopera, S., Krolikowski, J. (eds.) Innovation and knowledge in Innovative Enterprise, Wydawnictwo Politechniki Lodzkiej. Series: Monographs, Lodz (2007)
7. Zalewski, M.: Cisza w sieci Praktyczny przewodnik po pasywnym rozpoznawaniu i atakach posrednich. Helion, Warszawa (2005)
8. Wisniewska, M., Wisniewski, Z.: The relationship between knowledge security and the propagation of innovation. Adv. Intell. Syst. Comput. **783**, 176–184 (2019)

Beyond Passwords: Enforcing Username Security as the First Line of Defense

Thaier Fandakly and Nicholas Caporusso(✉)

Fort Hays State University, 600 Park Street, Hays 67601, USA
t_fandakly@mail.fhsu.edu, n_caporusso@fhsu.edu

Abstract. Combinations of account identifier (e.g., username) and key phrase (i.e., password) are among the most utilized form of credentials for several types of authentication purposes, such as, user verification, connection to public and private networks, and access to digital resources. Typically, usernames are considered a method of account or user identification, whereas passwords are regarded as the crucial component that protects from attackers and prevents breaches. As a result, the level of security of a set of digital credentials is primarily associated with the strength of the key phase, and most of the attention focused on promoting initiatives for increasing password security. Unfortunately, account identifiers received less consideration. Consequently, users are aware of how to enforce the security of their password, though they might prefer more convenient options. Contrarily, several bad practices are caused by overlooking usernames as the first line of defense. In this paper, we highlight the increasing importance of account names and we overview the main username practices that impact account security. Furthermore, we present the results of a study that evaluated how human factors and individuals' awareness impact username security.

Keywords: Authentication · Credentials · Cybersecurity · Password · Identity theft

1 Introduction

Access control and digital authentication algorithms typically use a combination of two types of information for authorizing a user: account identifier and password. In general, the former serves the purpose of finding a specific resource (or user) in the system, whereas the latter requires solving the challenge of knowing the secret word that grants access to the specific resource. To this end, the password is compared against the string associated with the identifier stored in the system. This method has been successfully utilized for decades for individual user accounts (e.g., e-mail mailbox, social media, and on-line banking accounts) as well as for accessing resources that can be shared among multiple users (e.g., public wireless networks).

Unfortunately, in the recent years, cyberattacks have become increasingly fierce thanks to the accessibility of information on the Internet, the availability of more powerful computational and network resources, and the development of more invasive techniques for threatening victims [1]. As an example, clusters of distributed devices,

T. Ahram and W. Karwowski (Eds.): AHFE 2019, AISC 960, pp. 48–58, 2020.
https://doi.org/10.1007/978-3-030-20488-4_5

that is, a botnet, can be utilized in a coordinated, brute-force attack to significantly reduce the time to crack (TTC), as discussed in [2]. Furthermore, as technology progresses and TTC decreases, more sophisticated key phrases are required in order to maintain a minimum level of security [3].

As a result, several organizations, such as, the National Institute of Standards and Technology (NIST), started issuing and updating guidelines for enforcing security of digital identities and for ensuring that specific assurance levels are met [4]. However, several factors prevent users from adopting best practices. On the one hand, lack of risk awareness and cybersecurity training is among the leading cause of weak credentials. Nevertheless, several studies showed that users tend to compromise security in favor of convenience. This, in turn, increases the success probability of attacks and the resulting damage, especially if the same information is reused among multiple different accounts or resources.

Although frameworks suggest a holistic approach to security, breach reports, cybersecurity guidelines, and awareness programs are primarily focused on increasing password security only; they give less attention to enforcing strong usernames. As a consequence, several malpractices regarding account identifiers are overlooked. For instance, users are provided with the opportunity of updating their passwords in most websites, whereas the possibility of changing account name is not considered, and, in some cases, not even allowed. Indeed, negligence in credentials management increases the impact of breaches: most recent cyberattacks are measured in millions of leaked credentials [5]. Moreover, some attacks address less-valuable targets with the purpose of gathering account credentials that can be exploited to access other platforms: in 2016, over 7 million accounts were leaked from a community on Minecraft, a popular virtual world especially utilized by youth for gaming and playful creativity [6]. Furthermore, hackers typically disseminate databases containing breached accounts in the dark web, which aggravates the problem: the so-called Anti Public Combo List contains more than 500 million usernames and passwords from various unrelated cyberattacks and data leaks that occurred over the past several years [7]. Simultaneously, this increases the risk and occurrence of credential stuffing, that is, the automated injection of breached username and password pairs on popular websites.

In this paper, we focus on the robustness of account names, and we present the results of a study in which we analyzed the human factors that affect security of credentials. Specifically, we were interested in evaluating whether technical background, security awareness, and familiarity with information technology have an influence on individuals' approaches to username generation strategies.

2 Related Work

A large body of scholarly work in cybersecurity mainly focused on passwords: studies showed that most users tend to create short and convenient key phrases that incorporate pieces of personal information, and that they reuse them across multiple accounts [8]. Therefore, in addition to stimulating awareness programs and guidelines for enforcing the security of passwords, research fostered the development of systems for generating more robust key phrases (e.g., requiring minimum length and special symbols) and more sophisticated access control procedures (e.g., two-factor authentication). Nowadays, users are provided with a considerable number of options for enforcing their password, though they might decide to deliberately ignore them.

In contrast, usernames received negligible attention: only a few studies specifically investigate their impact on account integrity [4, 5]. Nowadays, most Internet accounts use user's e-mail address or phone number as identification criteria, especially on popular web platforms (e.g., Facebook), whereas research focused on more personalized access credentials, such as, biometric identification. Nevertheless, regardless of the type of information utilized to recognize a user, account names are the first line of defense from a security standpoint, especially in websites that contain more valuable information (e.g., bank accounts, trading data), in personal devices, and in other private digital resources (e.g., routers and Wi-Fi connection).

The authors of [9] reported that very little attention has been given to the format of usernames and they concluded that most organizations use some variation on the first and last name of the user as an account identifier. This, in turn, makes usernames very easy to obtain. In [10], the authors studied that typically usernames reflect owner's habits and personal information. As for passwords, easy-to-understand usernames are convenient for individual user accounts even in presence of password management software [11]; moreover, identifiers are subject to the widely adopted practice of reusing the same name across multiple accounts [12].

3 Human Factors in Username Practices

The convenience of an easy-to-understand account name relies in the possibility of using strings that make it easier for users to identify a specific resource they want to get access to. This is particularly the case of open wireless networks (e.g., in airports, hotels, and cafeterias) for which a clear reference to the resource (e.g., AirportFreeWifi) makes it seamless for users to identify which network they can connect to. Indeed, this is suitable for assets that are meant to be easily discovered and utilized by individuals who potentially are not aware of the existence of said resource. In this case, the ergonomic aspect of an account name is a significant factor that justifies security trade-offs. On the contrary, user accounts, private wireless networks, and other digital assets that require higher degrees of privacy and access control should involve less compromise and guarantee security levels that are proportional to the sensitivity of information in a single account and to the potential risk that breaching one account has on other resources. In this regard, considering human factors is crucial for solving the optimization problem resulting from the inevitable trade-off between usability and complexity, that is, between effort to use and effort to crack.

In this Section, we review the key aspects that affect the security of account names: some of them (i.e., Sects. 3.1, 3.2, 3.3, and 3.4) are related to how access control systems are designed, whereas other factors (i.e., Sects. 3.5 and 3.6) are primarily determined by users' behavior. A third group of items consists of potential strategies that can be implemented as an improvement. Their technical characteristics and impact on ergonomics are detailed.

3.1 Account Names Shown in Clear

Access screens in personal computers, as well as registration and access forms on websites, typically show the account name in clear, whereas passwords are hidden or masked using characters that prevent individuals, including the user, from seeing them. Although password managers are designed to provide an additional security measure, they implement this mechanism as well [11]. This, in turn, might contribute to educating users that account names do not involve any security concerns and might lead to the misperception that account identifiers are expendable.

3.2 Social Exposure of Username

Indeed, risk is proportional to the number of people who have access to the account name and to the resource: while it can be lower for the credentials of a personal computers in a household, it is extremely higher in social media websites, where the account name (i.e., login identifier) and nickname are utilized interchangeably, openly shared with acquaintances, and exposed to billions of unknown users. Moreover, social media dynamics inherently promote advertising the account name in various ways as an instrument for gaining popularity and followers. Additionally, some platforms use the username itself as an access credential. In this context, the security of the account basically translates to the time required to crack the password only, as the username is already known. Although most social media websites have robust systems for preventing brute-force attacks, several hacking techniques could be utilized together to neutralize them.

3.3 Real-time Feedback About Duplicate Usernames

Some websites incorporate systems that facilitate account name selection by providing users with real-time feedback about their account identifier: as they type, the system notifies them if the name they chose is already taken, so that they can pick a different one. Although this streamlines account creation, it exposes existing account names to botnet attacks that could leak lists of usernames by attempting them. The impact of this breach might be significant because once account names are known, they can be exploited for further intrusion involving brute-force attacks to the password component of credentials, only. Indeed, such platforms might prevent this type of risk by introducing traffic-limiting and anti-bot systems. Nevertheless, given the probability of account name reuse, breached lists could be utilized to target other or resources that implement less-secure protocols against attackers.

3.4 E-mail Utilized as Account Name

As discussed earlier, nowadays most websites utilize e-mail addresses as account identifiers, which, in turn, might facilitate username breaches: e-mails are featured in company directories, on business cards, and among contact information in websites. Also, they are perceived as communication tools that can be advertised with no other risk than receiving more spam. Moreover, as it is very common in organizations to use standard e-mail formats (e.g., firstname.lastname@organization.website) [9], usernames are easy to guess.

3.5 Account Names Based on Personal Information

The dynamic described in Sect. 3.4 might induce individuals in the malpractice of reusing the same pattern (i.e., some combination of first and last name) as an account identifier for other resources or websites. As discussed in [13], the probability that two usernames refer to the same physical person strongly depends on the entropy of the username string itself. Experiments showed that weak usernames can be utilized to identify the same individual on different platforms and gather more information about their identity [9]. This is especially true when a single account name is reused. However, research showed that individual components of an identifier created using personal information (e.g., the nickname or last name) are enough to associate multiple accounts on different resources to a unique individual [12].

3.6 Account Name Reuse

Considering users' approach to password reuse, they might have a high tendency to reiterate the same identifier for other resources. As a result, both their username and key phrase are not unique. Furthermore, users' switching behavior results in keeping accounts open even if they are not being utilized [14]: information in the username can be used as a stepping stone to attack other services or websites that might have lower levels of security, for identity theft purposes [12]. The risks of the exposure of identifiers and passwords have been detailed by [15].

3.7 Strength Meters

Indeed, avoiding account reuse mitigates the impact of breaches. Conversely, having a more secure username reduces the risk of brute-force cyberattacks. In this regard, length and entropy are two fundamental features that determine the difficulty to guess a string, and thus, they are relevant to security, because they increase the time to crack [9]. Password meters utilize them for calculating the strength of key phrases and for providing users with feedback that helps them increase TTC. Unfortunately, studies showed that individuals approach password meters as a checklist and they do not enforce security beyond the minimum required level [16]. Using two strength meters (one for the identifier and one for the key phrase) might affect usability.

3.8 Enabling Account Name Change

In addition to creating a strong key phrase that maximizes TTC, changing password often is among the most effective practices for maintaining an account secure. This is already implemented in wireless networks and computer authentication. However, very few websites offer the option of modifying the account name, though most of them provide users with the option of changing their key phrase and even require users to update their password. Enabling the possibility of modifying their account names might help protect credentials. Indeed, this might disrupt the dynamics of websites, such as, social media, that use the account name as a nickname. Conversely, this opportunity could be utilized in other types of resources: to avoid impacting convenience, it could be given as an option, or requested ad hoc, whenever a major security concern arises.

3.9 Forcing Account Name Expiration

Many accounts, especially in the corporate world, have passwords that expire [17]. Username expiration could be forced in cases that require higher levels of security, such as, bank accounts. However, users might find it very inconvenient, though they are familiar with the procedure (i.e., similar to changing the password). Indeed, this would contribute to minimizing users' digital footprints, because accounts would automatically be deactivated upon expiration of the username. Nevertheless, many usability aspects make this option less actionable; as a remedy, it could be possible to define an expiration time that is inversely proportional to usage; this is to prevent deprecating accounts that are not utilized very often without having the possibility of alerting users; on the other hand, this would create some inconvenience for the users of resources that are frequently accessed, who would be required to change their username more often. Additionally, this solution might be beneficial from a privacy standpoint, as it would limit the number of resources that store users' information. On the other hand, websites would not accept this measure because it would affect their total user count, which is utilized by many companies to demonstrate large user bases, though they could use more accurate metrics (e.g., monthly active users).

4 Study

In the previous Section, we reviewed several aspects related to the security of a username and we discussed how human factors and ergonomics are involved in systems for account creation and access. Nevertheless, as the objective of our research is to improve the trade-off between effort to use and effort to crack, we realized a study to evaluate how users perceive the items described in Sect. 3 in terms of security and convenience of their account names.

Several datasets resulting from breaches enable investigating dynamics, such as, account name reuse. Conversely, our objective was to analyze: (1) whether users have a different approach to securing their identifiers depending on the type and importance of information in the account, and (2) the optimal trade-off between effort to use and effort to crack. To this end, we created a survey that asked respondents questions about their awareness and behavior in username generation in the context of several types of accounts requiring different levels of security.

A total of 120 participants (74 males and 46 females aged 31 ± 11) were recruited for this study: 20 subjects (G1) were sampled among people with a cybersecurity background or expertise, whereas the rest (G2) consisted of individuals with no specific training. By doing so, we utilized respondents with cybersecurity skills as a control group. Moreover, the survey consisted of questions that enabled us to compare the perception of username practices and password behavior.

5 Results and Discussion

In our analysis, we evaluated the aspects involved in the security and convenience trade-offs that are related to both system design and user behavior (i.e., Sects. 3.1, 3.2, 3.3, 3.4, 3.5, and 3.6). Results are reported in Table 1. Specifically, the answers of G1 were used as a gold standard to evaluate the difference in terms of perceived security between the two groups.

Conversely, in regard to convenience, we aggregated the data of the two groups: this skewed the results, though it had a negligible impact on the outcome of our analysis. This was for a two-fold reason: (1) although there was statistical significance ($p = 0.05$) between the groups for two of the considered dimensions, we found that perception of convenience was in general independent from cybersecurity training; also, (2) password generation and verification systems serve all users in the same fashion without considering their individual background or awareness.

Table 1. Results are shown in terms of perceived security (Sec), perceived convenience (Conv), and occurrence (Occ). As for the former two dimensions, a Likert scale (1 being the least secure/convenient) was utilized to collect responses; values regarding occurrence were directly collected in percentages. Data are reported separately for the control group (G1) and respondents with no cybersecurity training (G2).

Security factor	Sec. G1	Sec. G2	Conv.	Occ. G1	Occ. G2
Show the account name in clear (Sect. 3.1)	2.12	3.61	4.23	98.75%	100.00%
Use login information as account name (Sect. 3.2)	3.34	3.98	4.15	87.33%	78.22%
Give real-time feedback on duplicate username (Sect. 3.3)	3.53	4.62	3.11	8.44%	11.33%
Use the e-mail address as account name (Sect. 3.4)	2.44	4.32	4.54	68.80%	72.10%
Use personal information as account name (Sect. 3.5)	1.98	3.64	4.12	74.12%	79.78%
Reuse the same name for multiple accounts (Sect. 3.6)	1.52	3.37	4.27	82.22%	86.93%

Furthermore, we analyzed perceived security for the countermeasures outlined in Sect. 3 (i.e., Sects. 3.6, 3.7, 3.8, and 3.9) and we compared it with users' willingness to adopt them. Responses from the control group were considered as the gold standard in terms of perceived security, whereas the willingness to adopt was considered separately for each group. Table 2 and Fig. 1 show the results. In general, individuals with a cybersecurity background tended to be more open towards adopting mechanisms for protecting account names, though the difference between groups is not statistically significant ($p = 0.05$). Forcing account name expiration was perceived as affecting usability the most, and therefore this practice might be recommended only in websites that require higher protection standards.

Table 2. Perceived security (Sec) of prevention measures and willingness to adopt (WTA) them. As the control group was utilized as a gold standard, values for 3.1 through 3.6 are similar to the ones shown in Table 1.

Prevention measure	WTA G1	WTA G1	Sec. G1
Mask account name (Sect. 3.1)	76.11%	59.31%	2.12
Separate account name and nickname (Sect. 3.2)	68.40%	55.32%	3.34
Prevent real-time feedback on username (Sect. 3.3)	96.70%	84.21%	3.53
Avoid using e-mail as account name (Sect. 3.4)	65.66%	87.11%	2.44
Avoid using personal information in account name (Sect. 3.5)	74.12%	79.11%	1.98
Preventing account name reuse (Sect. 3.6)	56.12%	51.09%	1.52
Use strength meters (Sect. 3.7)	85.58%	78.29%	4.01
Enable account name change (Sect. 3.8)	96.63%	81.04%	4.23
Force account name expiration (Sect. 3.9)	8.20%	11.31%	4.45

Nevertheless, having the option of changing account name was accepted by the majority of respondents, and it received approximately 89% of preferences. Furthermore, users perceived that they would not be affected by removing real-time feedback about existing user names (i.e., Sect. 3.3), as this option was the most favored (a total of 90% on average among the two groups). However, this might be caused by lack of awareness of the consequences.

In general, users without cybersecurity training perceived current username practices as secure, and their responses show that they are more worried about their

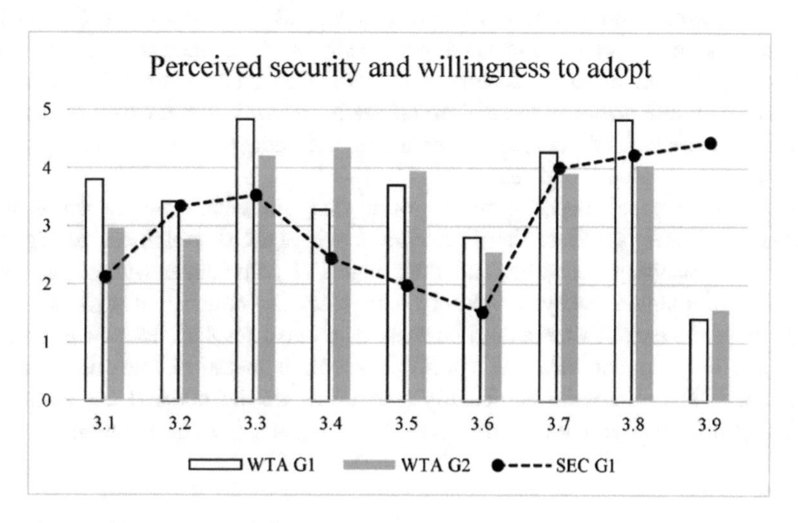

Fig. 1. Perceived security (line) and willingness to use (bars). Data about willingness to adopt (shown in Table 1) were converted from percentage to a Likert-equivalent to improve visualization. Labels on the X axis correspond to the security items outlined in Sect. 3 and described in Table 1.

passwords. This is consistent with the literature and with current practices. However, account leak due to real-time feedback was perceived as resulting in minimal risk by both G1 and G2. This practice was reported in 9.89% of cases, only.

Indeed, users know that reusing accounts that include personal information diminishes security, though both G1 and G2 reported doing it on average in 78.17% and 83.36% of the cases, respectively. This is consistent with findings in the literature [18], which report that users are aware of malpractices, but they prefer to more convenient options because they think they will not be affected by an attack. We did not detail account type because we found no statistically significant difference in perceived convenience and security.

6 Conclusion

Maintaining a secure account is the result of the continuous application of multiple strategies and practices that together contribute to preventing cyberattacks. In this paper, we focused on the importance of user credentials, and specifically, account names, in enforcing the security of accounts and authentication systems as the first line of defense. Indeed, there are alternatives to text-based account names: studies showed that graphical authentication might be safer than a long complex password that users might potentially forget. However, the authors of [19] highlighted that access control methods that contain graphical components are at high risk especially on mobile devices, because the authentication process can be followed on the screen. Moreover, although access methods based on biometrics [20] are increasingly being implemented in hardware devices, string-based account identifiers are still widely utilized in software and websites to enable individuals to log in.

We reviewed current practices in creating and using access credentials and we discussed the main issues associated with poor username security. In addition to highlighting the role of human factors in authentication systems, we outlined the risks caused by common practices and their implications in terms of user experience, and we detailed how the lack of strategies for enforcing username protection affects the trade-off between convenience and security.

Moreover, we reported the results of a study in which we evaluated users' awareness of best practices, their behavior and perceived usefulness in regard to methods for securing accounts, and their potential compliance with measures for improving username security. From our findings, we can conclude that most users do not perceive the lack of username robustness as a threat for their account information and, thus, they do not take any specific prevention measures, regardless of their background. The results are particularly relevant as we did not find any statistically significant difference in the case of accounts holding sensitive information.

References

1. Caporusso, N., Chea, S., Abukhaled, R.: A game-theoretical model of ransomware. In: International Conference on Applied Human Factors and Ergonomics, pp. 69–78. Springer, Cham, July 2018. https://doi.org/10.1007/978-3-319-94782-2_7
2. Dev, J.A.: Usage of botnets for high speed MD5 hash cracking. In: Third International Conference on Innovative Computing Technology (INTECH 2013), pp. 314–320. IEEE, August 2013
3. Brumen, B., Taneski, V.: Moore's curse on textual passwords. In: 2015 28th International Convention on Information and Communication Technology, Electronics and Microelectronics (MIPRO) (2015). https://doi.org/10.1109/MIPRO.2015.7160486
4. National Institute of Standards and Technology Special Publication 800-63B., p. 78, June 2017. https://doi.org/10.6028/NIST.SP.800-63b
5. Onaolapo, J., Mariconti, E., Stringhini, G.: What happens after you are pwnd: understanding the use of leaked webmail credentials in the wild. In: Proceedings of the 2016 Internet Measurement Conference, pp. 65–79. ACM, November 2016
6. Lenig, S., Caporusso, N.: Minecrafting virtual education. In: International Conference on Applied Human Factors and Ergonomics, pp. 275–282. Springer, Cham (2018). https://doi.org/10.1007/978-3-319-94619-1_27
7. Hunt, T.: Password reuse, credential stuffing and another billion records in have i been pwned, May 2017. https://www.troyhunt.com/password-reuse-credential-stuffing-and-another-1-billion-records-in-have-i-been-pwned/. Accessed 31 Jan 2018
8. Stainbrook, M., Caporusso, N.: Convenience or strength? Aiding optimal strategies in password generation. In: International Conference on Applied Human Factors and Ergonomics, pp. 23–32. Springer, Cham, July 2018. https://doi.org/10.1007/978-3-319-94782-2_3
9. Basta, A.: Computer Security and Penetration Testing, 2nd edn. Cengage Learning. VitalBook file (2015). Accessed 8 Aug 2013
10. Shi, Y.: A method of discriminating user's identity similarity based on username feature greedy matching. Paper Presented at the 2018 2nd International Conference on Cryptography, Security, and Privacy, March 2018. https://doi.org/10.1145/3199478.3199512
11. Wang, L., Li, Y., Sun, K.: Amnesia: a bilateral generative password manager. In: 2016 IEEE 36th International Conference on Distributed Computing Systems (ICDCS), pp. 313–322 (2016)
12. Jenkins, J.L., Grimes, M., Proudfoot, J., Lowry, P.B.: Improving password cybersecurity through inexpensive and minimally invasive means: detecting and deterring password reuse through keystroke-dynamics monitoring and just-in-time warnings. Inf. Technol. Dev. **20**(2), 196–213 (2013)
13. Perito, D., Castelluccia, C., Kaafar, M.A., Manils, P.: How unique and traceable are usernames? In: Privacy Enhancing Technologies. Lecture Notes in Computer Science, pp. 1–17 (2011). https://doi.org/10.1007/978-3-642-22263-4_1
14. Xiao, X., Caporusso, N.: Comparative evaluation of cyber migration factors in the current social media landscape. In: 2018 6th International Conference on Future Internet of Things and Cloud Workshops (FiCloudW), pp. 102–107. IEEE, August 2018. https://doi.org/10.1109/W-FiCloud.2018.00022
15. Thomas, K., Li, F., Zand, A., Barrett, J., Ranieri, J., Invernizzi, L., Bursztein, E.: Data breaches, phishing, or malware? Understanding the risks of stolen credentials. Paper Presented at the 2017 ACM SIGSAC Conference on Computer and Communications Security, October 2017. https://doi.org/10.1145/3133956.3134067

16. Caporusso, N., Stainbrook, M.: Comparative evaluation of security and convenience trade-offs in password generation aiding systems. In: International Conference on Applied Human Factors and Ergonomics. Springer, July 2019. (to be published)
17. Johansson, J.M., Brezinski, D.I., Hamer, K.L.: U.S. Patent No. US13277423, U.S. Patent and Trademark Office, Washington, D.C. (2011)
18. Tam, L., Glassman, M., Vandenwauver, M.: The psychology of password management: a tradeoff between security and convenience. Behav. Inf. Technol. **29**(3), 233–244 (2010). https://doi.org/10.1080/01449290903121386
19. Bošnjak, L., Brumen, B.: Improving the evaluation of shoulder surfing attacks. In: Proceedings of the 8th International Conference on Web Intelligence, Mining and Semantics (2018). https://doi.org/10.1145/3227609.3227687
20. Bevilacqua, V.: Retinal fundus biometric analysis for personal identifications. In: International Conference on Intelligent Computing, pp. 1229–1237. Springer, Heidelberg, September 2008

Social Engineering and the Value of Data: The Need of Specific Awareness Programs

Isabella Corradini[1,3(✉)] and Enrico Nardelli[2,3]

[1] Themis Research Center, Rome, Italy
isabellacorradini@themiscrime.com
[2] Department of Mathematics, Univ. Roma Tor Vergata, Rome, Italy
nardelli@mat.uniroma2.it
[3] Link&Think Research Lab, Rome, Italy

Abstract. In the field of cybersecurity human factor is considered one of the most critical elements. Security experts know well the importance of people's security behaviors such as managing passwords, avoiding phishing attacks and similar. However, organizations still lack a strong cybersecurity culture to manage security risks related in particular to the human factor. In this paper we describe the results of a study involving 212 employees belonging to two companies operating in the service sector. Within a cybersecurity awareness project executed in each company, employees participated in workshop sessions and were asked to evaluate the credibility and the success probability of a list of the most common security risk scenarios based on social engineering techniques. Cyber-attacks based on these techniques are considered among the most successful because use psychological principles to manipulate people's perception and obtain valuable information. The comparison of results obtained in the two companies shows that awareness training programs pay off in terms of raising people's attention to cyber-risks.

Keywords: Human factors · Cybersecurity · Social engineering · Cyber hygiene · Awareness

1 Introduction

Cybersecurity is a hotly debated topic all over the world, and the protection of information is a priority for institutions, companies and individuals. A data breach can have a high financial impact on a company, considering that in the range of 1 million to 50 million records lost, breaches can cost companies between $40 million and $350 million respectively [1]. In addition, companies have also to consider other significant consequences for their business, such as the loss of intellectual property and reputational damage [2].

Cyber threats have been growing over the last few years and they are going to be based on the exploitation of new opportunities.

On the one hand, security international reports stress the impact of the old cyber threats, such as ransomware, phishing, spear phishing, data breaches (e.g. [3–5]). Moreover, they highlight the importance of human factor, since mail and phishing

T. Ahram and W. Karwowski (Eds.): AHFE 2019, AISC 960, pp. 59–65, 2020.
https://doi.org/10.1007/978-3-030-20488-4_6

represent the primary malware infection vector [3] while social engineering is a critical launchpad for email attacks [5].

On the other hand, new threats are made possible by the application of Internet of Things and Artificial Intelligence [6]; furthermore, these technologies can strengthen the existing threats, such as improving the frequency of phishing attacks.

Notwithstanding the fact that more and more innovative technical solutions are available on the market to provide protection to companies and institutions, the problem of cybersecurity is far from being solved.

The role of human factor in cybersecurity is a fundamental topic to gain a better defense against cyber-attacks. Many authors indeed stress the importance of adopting a holistic approach, given that cyber defense cannot be considered only from a technical perspective but requires also a human-social viewpoint (e.g. [7–9]).

This paper is focused on workers' perception of cyber-attacks based on social engineering (SE), which is a method using psychological principles to manipulate people's perception to gain their confidence and lead them to disclose sensitive information or to do something else (e.g. opening an e-mail attachment), for the benefits of those who use these strategies (e.g. [8, 10]). SE is a successful technique because it exploits human nature bypassing technological measures [11]. In fact, as reported in [10], «We, as human beings, are all vulnerable to being deceived because people can misplace their trust if manipulated in certain ways».

SE can be used for several purposes and by different actors, targeting people through information directly posted by Internet users. SE can be executed in different forms. *Phishing*, a massive distribution of emails to solicit personal information, and *spear phishing*, targeting victims individually, are a form of SE. Moreover, SE can exploit physical devices (*baiting*), for example an infected USB stick left unattended in order to be found and used by people, with the consequence of installing malware onto the computer. Finally, SE can be executed by phone (*vishing*) to trick people or by exploiting information collected during a *face to face conversation*. Even though the actual modalities of execution can cause different reactions in people [12], the focus of SE is the social interaction.

2 Methodology

The study has involved 212 employees belonging to companies operating in the service sector (94 in company X, and 118 in company Y). In each company, we have carried out a cybersecurity awareness project aimed at the building of security culture. We used an interactive approach to actively involve participants and discuss with them security problems, and how to manage them.

More specifically, within each project we gathered 3–4 groups belonging to the same company for a half-day workshop where we tackled some of the most common security risk scenarios related to human behavior (e.g. choosing secure password, using unsecure wi-fi services). We repeated this half-day workshop with different sets of groups until all involved employees had attended. There were in total 13 groups for company X and 16 for company Y, with an average of 7 per group.

In each workshop, group participants were presented with the list of considered security risk scenarios and were asked to assign a mark to the **credibility** of each of them (i.e., how plausible the scenario is) and to its **success probability**, using a scale from 1 (low) to 5 (high).

At the beginning of each workshop we explained, to all groups present, each of these security risk scenarios, by showing videos in the public domain or short excerpts from well-known movies depicting the specific scenario and by illustrating real life examples of them (e.g. actual phishing emails). Subsequently, groups split and each of them separately discussed the presented scenarios, in order to estimate its credibility and success probability in the light of their personal experience, both in business and in private life.

After each group internally discussed and provided a consensus evaluation on both the credibility and the success probability of the scenarios, we united all groups together and a representative from each of them declared their conclusion. Next, we conducted a discussion and a comparison among all participants in that workshop of the various conclusions. Finally, we trained participants on the best behavioral practices to manage the presented security risk scenarios.

Some of these security risk scenarios were based on social attacks and engineering techniques (e.g. phishing), still a relevant problem given that social attacks are very frequent and can compromise data, secrets, and credentials [4]. The security risk scenarios discussed in the paper are the following:

- Receiving emails asking for data or to perform some action (Phishing and spear phishing)
- Receiving a phone call asking for information (Vishing)
- USB dropped in obvious places to employees (USB baiting)
- Face to face conversation

Note that the first three above listed scenarios refer to situations that intrinsically constitute direct risk scenarios, in the sense that they directly lead to jeopardize valuable assets. On the other side the last scenario describes a situation where there is not an immediate danger but the consequences of careless behaviors may provide a social engineering attacker with the information on which to successfully carry out the above three scenarios.

3 Results and Discussion

We now report and discuss the main outcomes of our study related to the above listed scenarios, also in the light of the different situations existing in companies X and Y.

In Figs. 1 and 2 we compare credibility and success probability results obtained in each of the two companies. Reported numbers are, for each risk scenario, the average across all groups involved of the consensus evaluation provided by each group.

As you can see, the success probability has an average mark slightly lower than credibility in all scenarios apart from "Vishing" in both companies. This scenario refers to a kind of interaction where people are naturally aware of the risk of being unconsciously manipulated by astute people. Even without training or previous experience, it

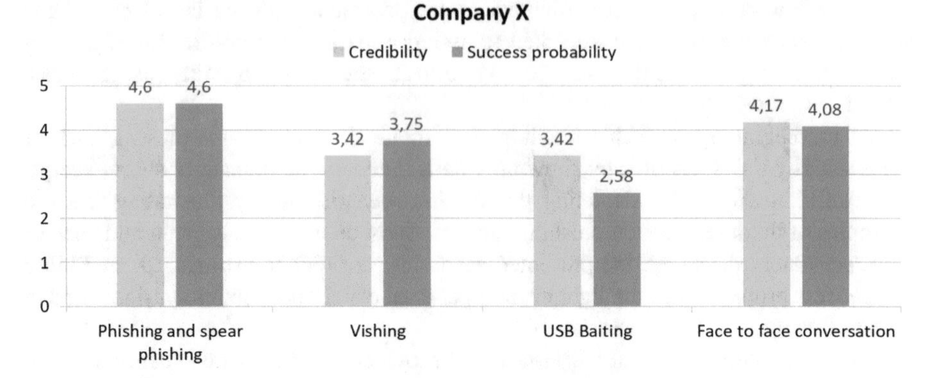

Fig. 1 Results for Company X.

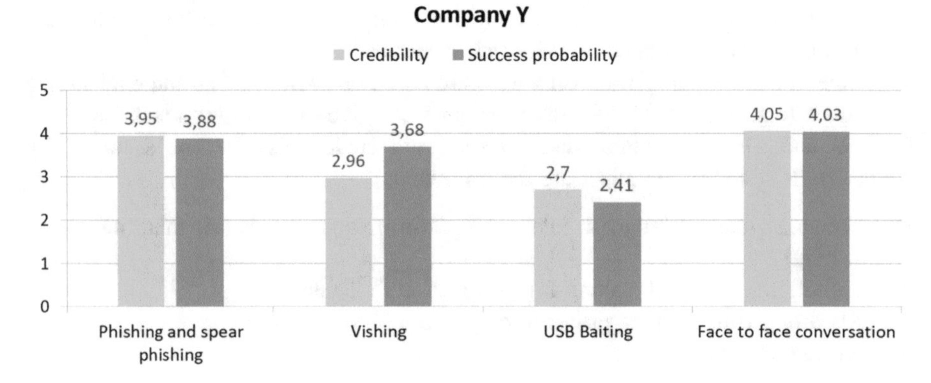

Fig. 2. Results for Company Y.

appears highly plausible to many that an able and empathic speaker can persuade others during a phone conversation.

Also, scenario "USB Baiting" has in both companies the lower mark, most probably because the specific situation where a memory stick is dropped in obvious places for employees is not a common happening. Moreover, it depends on the security policy adopted by organizations, given that the use of a USB stick could be prohibited.

Finally, scenario "Face to face conversation" has received the highest mark in one company and the second highest in the other one, which is reasonable given that face to face interactions are common in any kind of job and people are aware that these situations can be a very good opportunity to collect sensitive information.

In Figs. 3 and 4 we present the same data but arranged to compare the situation between the two companies.

Figure 3 presents credibility marks. You can see that, in general, employees in Company Y are less convinced by the plausibility of the presented risk scenario than in Company Y. This may be explained by the fact that company X has been working on a security culture project for a few years and their employees have been participating in

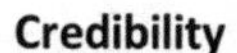

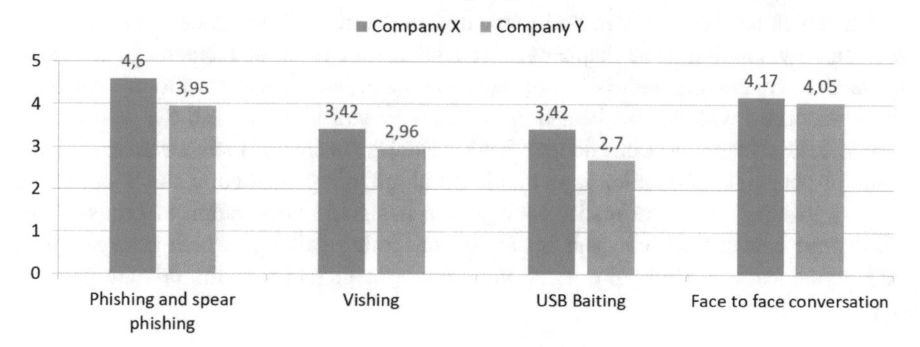

Fig. 3. Results for credibility.

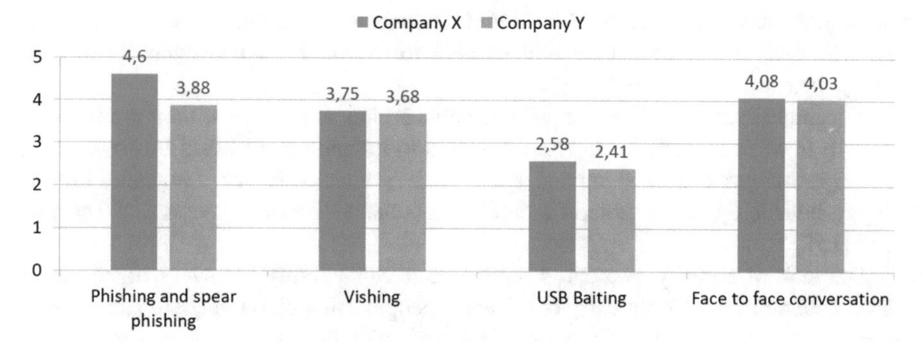

Fig. 4. Results for success probability.

specific training sessions, while company Y is at its first experience. Moreover, most participants of the company Y are not aware that an email can be sent to targeted individuals such as spear phishing, as well they were not aware of the psychological aspects of these security risks.

Moreover, as reported by periodical reports on the most common types of cyber-security threats and cited above, phishing is actually the preferred vehicle for SE attacks.

For what regards the success probability, whose comparison of the marks between the two companies is shown in Fig. 4, there is no published data – to the best of authors' knowledge – about the actual success rate of the various threats. Annual cybersecurity reports usually provide indications on the top threats and whether a threat has become more or less common compared to the previous year [4, 5, 15]. This is understandable since data about failed attacks are usually not disclosed, while successful attacks cannot be usually hidden, for both their visible consequences and data protection laws (e.g. GDPR, General Data Protection Regulation) that requires companies to notify an authority data breaches when they occur.

Another important aspect concerns data protection related to the use of social media, which has been tackled during the discussion of the "face to face conversation" scenario, by relating what happens on social media to what happens in a person interaction. Discussing their behavior on social media, employees tend to minimize the risks associated with certain behaviors. From their words it emerged that while company's data protection is considered fundamental to business, their awareness of the value of personal information is not so high: an often repeated comment was "I have nothing to hide", while in reality each person has some information to protect. This dichotomy between attitude and behavior concerning privacy, which emerged with higher frequency in Company Y, is well-known in literature as the privacy paradox (e.g. [13, 14]).

4 Conclusions

It is clear that while digital technology is spreading everywhere security risks are growing and have to be seriously tackled. Criminals tend to exploit every vulnerability they can find; in addition, they will be able to exploit the advantages of Artificial Intelligence and Internet of Things.

If technical solutions are adequate to solve technical problems, they are inappropriate to manage security cyber threats related to human nature based on social engineering technique, e.g. phishing and spear phishing attacks. Hence, companies have to adopt a holistic approach, able to include and balance "People, Process and Technology" [15].

The lack of security awareness represents a vulnerability for every organization, making SE attacks easier to carry out. Hence, people using digital technologies have to be more and more aware of the risks involved with their use. In fact, even though cybersecurity is considered by governments and institutions as a priority, the actual behavior of people represents a challenge for any organization [16].

Therefore, building a cybersecurity culture in organizations [17, 18] is the best way to develop and reinforce effective security practices [19].

In this paper we have described the outcome of a study involving 212 employees, belonging to two companies in the service sector, who participated to a cybersecurity awareness project aimed at the building of a security culture within the organization. Employees had to evaluate the credibility and the success probability of each security risk scenario presented.

In one company the project was carried out for the first time, while in the other people had already participated in cybersecurity awareness training sessions. The analysis shows that people in the latter company have a better comprehension of risks related to the use of digital technologies.

Our study therefore provides support for the fact that without adding people to a company defense arsenal, effectiveness of its cybersecurity is weakened. This is in line with recommendations of recent cybersecurity reports [3, 5].

References

1. Ponemon Institute: Cost of a Data Breach Study: Global Overview (2018). https://databreachcalculator.mybluemix.net/assets/2018_Global_Cost_of_a_Data_Breach_Report.pdf
2. Allianz: Allianz Risk Barometer. Top Business Risks for (2018). https://www.agcs.allianz.com/assets/PDFs/Reports/Allianz_Risk_Barometer_2018_EN.pdf
3. ENISA: Threat Landscape Report. 15 Top Cyberthreats and Trends (2018). https://www.enisa.europa.eu/publications/enisa-threat-landscape-report-2018
4. Verizon: Data Breach Investigation Report (2018). https://enterprise.verizon.com/resources/reports/DBIR_2018_Report_execsummary.pdf
5. CISCO, Cisco 2018 Annual Security Report (2018). https://www.cisco.com/c/dam/m/digital/elq-cmcglobal/witb/acr2018/acr2018final.pdf
6. Brundage, M., Avin, S., Clark, J., et al.: The malicious use of artificial intelligence: forecasting, prevention, and mitigation (2018). https://arxiv.org/abs/1802.07228
7. Schultz, E.: The human factor in security. Comput. Secur. **24**(6), 425–426 (2005)
8. Corradini, I.: Human factors in hybrid threats: the need for an integrated view. In: Zorzino, G., et al. (eds.) Hybrid Cyberwarfare and The Evolution of Aerospace Power: Risks and Opportunities, pp. 85–96, CESMA (2017)
9. Ki-Aries, D., Faily, S.: Persona-centred information security awareness. Comput. Secur. **70**, 663–674 (2017)
10. Mitnick, K.D., Simon, W.L.: The Art of Deception: Controlling the Human Element of Security. Wiley, New York (2002)
11. Schneier, B.: Secrets and Lies. Wiley, New York (2000)
12. Bullée, J.W.H., Montoya, L., Pieters, W., Junger, M., Hartel, P.: On the anatomy of social engineering attacks: a literature-based dissection of successful attacks. J. Invest. Psychol. Offender Profiling **15**(1), 20–45 (2018)
13. Barnes, S.: A privacy paradox: social networking in the United States. First Monday, **11**(9) (2006). https://firstmonday.org/article/view/1394/1312_2
14. Barth, S., de Jong, M.D.T.: The privacy paradox: investigating discrepancies between expressed privacy concerns and actual online behavior – a systematic literature review. Telematics Inform. **34**(7), 1038–1058 (2017)
15. Schneier, B.: https://www.schneier.com/blog/archives/2013/01/people_process.html
16. De Bruijn, H., Janssen, M.: Building cybersecurity awareness: the need for evidence-based framing strategies. Gov. Inf. Q. **34**, 1–7 (2017)
17. Enisa: Cyber Security Culture in organizations (2018). https://www.enisa.europa.eu/publications/cyber-security-culture-in-organisations
18. Corradini, I., Nardelli, E.: Building organizational risk culture in cyber security: the role of human factors. In: AHFE 2018, pp. 193–202. Springer, Cham (2018)
19. Wilson, M., Hash, J.: Building an information technology security awareness and training program. NIST Special Publication 800-50, USA (2003)

Awareness and Cyber-Physical Security

Human Centered Cyber Situation Awareness

Vincent Mancuso(✉), Sarah McGuire, and Diane Staheli

MIT Lincoln Laboratory, 244 Wood Street, Lexington, MA 02421, USA
{vincent.mancuso, sarah.mcguire, diane.staheli}@ll.mit.edu

Abstract. Cyber SA is described as the current and predictive knowledge of cyberspace in relation to the Network, Missions and Threats across friendly, neutral and adversary forces. While this model provides a good high-level understanding of Cyber SA, it does not contain actionable information to help inform the development of capabilities to improve SA. In this paper, we present a systematic, human-centered process that uses a card sort methodology to understand and conceptualize Senior Leader Cyber SA requirements. From the data collected, we were able to build a hierarchy of high- and low- priority Cyber SA information, as well as uncover items that represent high levels of disagreement with and across organizations. The findings of this study serve as a first step in developing a better understanding of what Cyber SA means to Senior Leaders, and can inform the development of future capabilities to improve their SA and Mission Performance.

Keywords: Cybersecurity · Situational awareness · Card sort

1 Introduction

Situation Awareness (SA) is an important antecedent to improved individual and team performance in complex environments. Originally studied within the context of aviation, SA has since been extended to numerous domains such as air traffic control [1–3], emergency response operations [4, 5] and intelligence analysis [6], amongst others. Situation Awareness is defined as "the perception of elements in the environment with a volume of time and space, the comprehension of their meaning and projection of their status in the near future" [7]. This model implies that SA is a state of human cognition, in which a decision maker uses a separate perception-action cycle that utilizes a mental model to assess the current situation, and project different actions against the environment.

More recently, SA has become identified as a critical component within cybersecurity and operations. From a military perspective, Joint Publication 3–12, Cyber Space Operations [8], defines Cyber SA as "The requisite current and predictive knowledge of cyberspace and the operating environment upon which cyberspace operations depend,

Distribution Statement A. Approved for public release. Distribution is unlimited. This material is based upon work supported under Air Force Contract No. FA8702-15-D-0001. Any opinions, findings, conclusions or recommendations expressed in this material are those of the author(s) and do not necessarily reflect the views of the U.S. Air Force.

T. Ahram and W. Karwowski (Eds.): AHFE 2019, AISC 960, pp. 69–78, 2020.
https://doi.org/10.1007/978-3-030-20488-4_7

including all factors affecting friendly and adversary forces", and generalizes to three aspects of awareness, ***Network Awareness***, ***Mission Awareness***, and ***Threat Awareness***. These areas of awareness can be further divided based on the activities of ***friendly (blue)***, ***neutral (gray)*** and ***adversary (red)*** forces and their capabilities across the spectrum of conflict. Cyber SA for cyber defense has been described as containing seven critical aspects of awareness; the current situation, the impact of the attack, the evolution of the situation, adversary behavior, the causation of the situation, the quality of the information, and the plausible futures of the current situation [9].

Unlike many of the environments in which SA has traditionally been studied, cyberspace has additional complexities due to its dynamic and intangible nature. Because the environment itself does not have a physical form with meaning, cybersecurity analysts often rely on technology to form, maintain and utilize their Cyber SA, often referred to as capabilities. Cyber SA capabilities produce a significant amount of data about the environment and its current state. If not curated and designed correctly, this leads to a gap between the available data and the data that is necessary in the current task, also known as the decision-making gap [10]. Rather than focusing on delivering a deluge of data, Cyber SA capabilities should focus on understanding the situation, and providing the necessary information in a way that is cognitively usable and facilitates the human decision-making process.

To best augment SA and thus human performance, success capabilities must take into account the ***people***, ***processes,*** and ***technology*** needed. Previous models of SA suggest that it is a human cognitive state that can be assisted by technology, and distributed across organizations and teams that have concrete tasks to complete. Therefore, depending on their job role, organization and mission, the requisite SA information and processes around them, will vary across the different user groups in cyber operations. From a capability perspective, this implies that Cyber SA tools must provide accurate, timely, comprehensive and actionable information for a wide spectrum of possibilities, user groups, and tasks. When developing capabilities, this creates two necessary questions that must be answered for each user group, (1) How do you characterize different user groups across cyber operations and (2) What are the Cyber SA information requirements for each of those user groups.

1.1 Purpose

The purpose of this research was to develop a holistic view of Cyber SA of Senior Leaders across different organizations in cybersecurity. We leveraged a card sort methodology [11, 12] to extract the critical Cyber SA requirements. The card sort task provided us with data on how Senior Leaders prioritize information allowing us to build a hierarchy of high- and low- priority Cyber SA information, as well as uncover several shared schemas that can be used to drive capability development and serve as boundary objects for collaboration across mission partners. The goal of this paper is to not only provide empirical direction towards the development of Cyber SA capabilities focused on specific information requirements, but to demonstrate an alternate methodology for user-centered design in developing Cyber SA technologies.

2 Method

2.1 Materials

Using the [Network, Mission, Threat] × [Blue, Grey, Red] paradigms, we developed a set of 69 Cyber SA information categories that were refined amongst the research team and other Subject Matter Experts. Table 1 shows the general categories of cards across the 3 × 3 paradigm space. It is important to note that these Cyber SA information categories, were designed to represent the information needs of Senior Leaders, and not necessarily the entire problem space of Cyber SA, thus the strong skew towards Blue and Mission related information.

Table 1. Categories of cards within [Network, Mission, Threat] × [Blue, Grey, Red] paradigm

	Red	Grey	Blue
Network	Adversary Key Cyber Terrain (1 Card)	Partner Key Terrain and Security Compliance (2 Cards)	Blue Key Terrain, Security Compliance and Current Cyber/Network Activity (17 Cards)
Mission	Top Threat Groups/Actors and Associated TTPs (2 Cards)	Partner Missions and Associated Risk, Relevant Geo-Political News (5 Cards)	Blue Mission, Staffing, and Readiness (26 Cards)
Threat	Known Threats and Vulnerabilities of Adversaries (2 Cards)	Top Threat Activity, Vulnerabilities, and Incidents Affecting Partners, and Global Security and Attacks (6 Cards)	Top Threat Activity Affecting DoD & US, and this Organization and Indicators of Insider Threat (8 Cards)

After crafting an initial set of Cyber SA categorizations, we piloted the study with participants familiar with network security. During these pilots, we adjusted the language of the cards, added new cards and refined the methodology to be used during exercises with Senior Leaders. Once the final set of relevant Cyber SA information was agreed upon, cards were printed on laminated 2.5 × 3.5 cards (the size of a playing card) with a unique ID number in the upper left-hand corner for the purpose of logging (Fig. 1).

Fig. 1. Sample cards used for Cyber SA Elicitation Card Sort exercise

2.2 Participants and Procedure

Card sort activities were performed with 19 individuals (Table 2). The individuals spanned multiple organizations, commands, and echelons of leaderships. Results are presented by organization, which were defined as Executive Leadership, Intelligence, and Operations. Due to the low number of participants from Intelligence for the majority of the analysis the data for Intelligence and Operations were combined.

Table 2. Study participants by echelon and organization.

	Executive Leadership	Intelligence	Operations	Total
Lt. Colonel/Commander	1	0	3	**4**
Colonel/Captain	0	1	3	**4**
Brig General/Rear Admiral	0	1	1	**2**
Maj General/Rear Admiral	3	0	1	**4**
Lt General/Vice Admiral	1	0	0	**1**
Civilian/SES	3	1	0	**4**
Total	**8**	**3**	**8**	**19**

At the beginning of each activity Senior Leaders were briefed on the intended goal of the study. They were then presented the set of cards and received instructions. Senior Leaders were asked to sort the cards based on priority. They were asked to create three piles consisting of high, medium, and low priority items. High priority items were defined as information that is crucial to their job and that they want to see on a frequent basis. Medium priority items were defined as information that they want to see but not on a frequent basis. Low priority items were defined as items they do not need to see very often if at all. The size of each of the three groupings could be as large or as small as they wanted. Blank cards were available during the exercise if they felt a type of information was missing from the deck. Once all cards were sorted, the researcher inquired into why the participant organized the cards the way they did. During this phase, participants were given freedom to rearrange cards.

3 Results

Overall a high variance in the priority ratings across Senior Leaders was found. However, analyses including descriptive statistics, overall prioritization rankings, and inter- and intra- organization disagreement in rankings showed some interesting patterns for discussion.

3.1 Descriptive Statistics

The data revealed that overall participants rated more cards as High priority (M = 25.63, SD = 11.10) than Medium (M = 24.11, SD = 13.11), and both more than Low (M = 14.84, SD = 14.35). Executive Leadership had a higher priority skew (M = 26.57, SD = 8.17), with a more similar count for Medium (M = 23.00, SD = 12.74) and Low (M = 19.43, SD = 12.74). Intelligence and Operations had a more equal distribution between High (M = 26.6, SD = 11.9) and Medium (M = 27.2, SD = 12.84) with fewer items rated as Low priority (M = 14.7, SD = 14.33).

3.2 Overall Prioritization Rankings

Cards were summed across all participants and normalized based on the number of participants. The results showed that there were no Cyber SA information categories which had unanimous agreement, either high, medium or low priority. There were five items which 13 or more participants agreed were high priority (Table 3).

Table 3. High priority Cyber SA items

Card	Low	Med	High
Indications of insider threat - this organization	2	2	14
Top threat activity affecting DoD & US	0	2	14
Top cyber-attacks - this organization	1	4	13
Status of Operations (Current, Next 24 h, Past 24 h)	2	3	13
Battle damage assessments conducted by this organization	2	3	13

Similarly, there were five items which were agreed as being of lowest priority, with only one participant indicating that they were high, and 9 or more suggesting they were low priority Table 4.

Table 4. Low priority Cyber SA items

Card	Low	Med	High
Status of service desk tickets - this organization	15	2	1
Upcoming events (e.g. inspections) - blue force	12	4	1
Security hygiene - partners	12	4	1
Status of ongoing security initiatives - this organization	11	6	1
Security compliance - partners	9	7	1

3.3 Across Organization Disagreement

Overall priorities for Executive Leadership and Intelligence and Operations were calculated. Cards were scored within each participant based on whether they were ranked as high (receiving a 2), medium (1) or low (0) and then a summative score was calculated for each card. Since there was a differing number of participants across groups, a normalized score was calculated for each card based on the number of raters. From the normalized scores, elements that represented differing organizational priorities were extracted by calculating the difference in priorities (subtracting the normalized scores).

The results showed that Executive Leadership had a higher value on their partner status, with higher rankings for Partner Key Cyber Terrain (8.8) and Risk to Key Assets (12.5) than Intelligence and Operations (5.5 and 7.7). On the other hand, the Intelligence and Operations held higher value on information on their organizations cyber posture and status, with higher rankings for Global Security Alert Levels (15.0) and Security Compliance and Hygiene (13.9), than Executive Leadership (7.5 and 4.4).

3.4 Within Organization Disagreement

In addition to looking at the scores across organizations, agreement ratings within each organization were calculated to better understand where there was the most disagreement. Agreement was calculated based on methods for calculating interrater reliability as proposed by Fleiss [13]. The value, which ranges from 0 (no agreement) to 1 (complete agreement) represents the extent to which the participants agree for a given card (i.e. how many participant-participant pairs are in agreement on the rating, relative to the number of all possible pairs).

For Executive Leadership, the lowest agreement was for the Key Blue Force Assets of Partners (25.00% agreement) and Potential Threat to Partners Key Cyber Terrain (32.1%). For Intelligence and Operations Top Cyber Incidents of Partners (27.78%) and Risk to Partner Key Assets (27.78%) had the highest disagreement.

4 Discussion

As part of this study we wanted to gain a better understanding of the Cyber SA information requirements of Senior Leaders. A high agreement on what constituted high and low priority was found. Across individuals it was found that items related to threat activity including Top Cyber Attacks and Indication of Insider Threat were considered to be of highest priority. These items have direct impact to mission and require immediate defensive steps to be taken to reduce vulnerabilities, block potential threats, and take other remediation steps. All of the top priority items additionally referred to the Senior Leader's own organization and not partner organizations, implying that Senior Leaders are inward thinking when it comes to their Cyber SA information needs. The lowest priority items were system administrative items including Security Compliance and Security Hygiene of partner organizations and Status of Service Desk Tickets for their own organization. While important for securing the network, this information is perhaps too granular for Senior Leaders as it has an unclear direct association to ongoing missions.

While there was agreement on what constituted high and low priority, there was a high level of ambiguity on what information was considered of medium priority. While there are multiple potential reasons for this, one thing that was noticed while running these card sort activities, was the variance in interpretation for many of these Cyber SA concepts. During the sorting process, the researchers provided little explanation, allowing the participants to interpret the items themselves. The highly variable interpretations indicate that Senior Leaders have different expectations and understandings of the meaning of many of these concepts.

Another interesting finding, was in the agreement on priorities across the organization. Based on their organizational missions, it was expected that there would be some disagreement in how each organization prioritized information. Our findings showed that Executive Leadership had a higher value on information about partner organizations, while Intelligence and Operations had more interest in information on the cyber posture and status. Executive leadership is responsible for maintaining awareness of the enduring mission scope, in which the on-going missions are situated. The interaction of enduring missions and the entire set of current ongoing missions involve a broader scope, and have dependencies on external entities. This is evident in them considering key terrain and associated risk to the partner organizations to have higher priority. Missions often have external dependencies with partner organizations, and the success or failure of those missions has dependencies with the partner's cyber status.

Unlike the Executive Leadership, Senior Leaders in Intelligence and Operations often have a singular (or small number) mission focus. From this perspective, they have a much more fine-grained viewpoint of individual mission dependencies within their organization. Participants in Intelligence and Operations held higher regard to information on the local organizations Security Compliance and Hygiene, and the Global Security Alert Levels. This information is important as it provides Intelligence and Operations with information on what the current readiness and posture of their current organization is at any given point and time, and how that information may interact with the security alert levels that are being reported across the broader security community. This information is critical in helping them better focus their mission directives, and achieve the necessary effectiveness.

The final discussion item was based on the disagreement within the organizations. Within both Executive Leadership and Intelligence and Operations, there were a few items that had a surprisingly high level of disagreement across participants. Within both organizations there was disagreement on the role of partners in their Cyber SA. Executive Leadership, had high levels of disagreement on information on the Key Blue Force Assets and Threats to Key Terrain of their partners, while Intelligence and Operations had disagreement for Partners Top Cyber Incidents and the Risk to Key Assets. These disagreements, resonate with the differing priorities discussed earlier. For Executive Leadership, with a broader mission focus, information on Blue Force Assets and Threats to Key Terrain is very important with participants agreeing that those items were either high or medium priority for their own organization, however there was disagreement in priority rankings for partners. Similarly, Intelligence and Operations had disagreements on Cyber Incidents and Risk to Key Assets, however for those items focused on their organization, there was much higher agreement and prioritization. We feel that these findings indicate a continuing ambiguity on how partner infrastructure and status impact the local mission.

5 Implications

The purpose of this work was the help better inform the development of capabilities for Senior Leaders Cyber SA, in addition to practical implications, there are also numerous research implications.

It was hypothesized before conducting the card sorts that a large number of Cyber SA items would be rated of low priority and a small number would be rated of high priority. However, on average Senior Leaders rated more than 30% of the Cyber SA information items as High priority. Senior Leaders receive their awareness through some sort of regular update brief (in military organizations these are often referred to as Commanders Update Briefs, CUB). These update briefs are often a few PowerPoint charts, and reflect what their staff think represents the Commanders Critical Information Requirements (CCIRs). However, with the large amount of High priority Cyber SA information it is unlikely they are receiving the awareness that they desire. This implies that there is research needed to help develop capabilities to help Senior Leaders better identify and communicate their CCIRs, and enable the communication pipeline of those requirements from their staff.

One of the biggest surprises in the analysis was that there was very little overlap or uniformity in the prioritizations, even within an organization. This finding indicates that Cyber SA may not be generalizable rather, Cyber SA is driven by context, specifically mission priorities, current needs and risk tolerance. From a research standpoint, this implies that further work is needed in order to map how the context of the mission and current status of an organization impacts Cyber SA needs. From a design standpoint, Cyber SA capabilities must be designed to be composable, to allow Senior Leaders and their staff to customize which information they see, and modify it based on their current priorities, needs, and mission focus.

There was also high ambiguity in the rating of medium priority items, which maybe due to different interpretations of Cyber SA concepts. This could result in diverging mental models, and lack of common ground, which could become a barrier in collaboration across organizations. Further research on operationalizing and defining fuzzy Cyber terms may help improve agreement, and thus collaboration across organizations. Additionally, from a capability development perspective, focus can be put on developing capabilities that act as boundary objects [14] that help translate understanding of various Cyber SA concepts across organizational boundaries. These capabilities would help link differing understandings, and expectations of Cyber SA to relevant data, and serve as a collaborative artifact to improve communication, collaboration and decision making.

6 Conclusions and Future Work

In this paper, we aimed to better understand how Cyber SA information is prioritized by Senior Leaders. Using a card sort methodology, we collected data from 19 Cyber Operations Senior Leaders, to elicit their own individual Cyber SA needs, and identify commonalities of high-priority missions. Our findings showed that information focused on current threats and attacks against the organization had the highest prioritization across participants, while system administration information was the lowest. The findings also showed that participants in Executive Leadership held higher value on information on partner status, while Intelligence and Operations valued information on their own organizations cyber posture and status. Additionally, across all participants, there was a high amount of disagreement on the prioritization of partner nations status.

While the data and findings we collected in this study are insightful, continued work is necessary in developing a more holistic understanding of Cyber SA. For this study, we focused singularly on Senior Leaders, however they only account for a portion of the decision making in cyber operations. Future work should engage with personnel at the tactical, and operational levels, drawing from analysts, planners, and operators, to gain insight into how Cyber SA differs across work-roles, and to identify boundary objects for collaboration across an organization.

For this study, we developed materials that had a heavy skew towards defense, with a heavy blue mission focus, this decision was made based on the target demographic for participants. Future work should expand upon the cards to account for other types of operations (Offensive and DoDIN Ops), and provide more coverage of the Network, Threat and Red and Grey space. Coupled with an expanded participant demographic, the introduction of this information can provide a more complete understanding of Cyber SA prioritizations.

Finally, the card sort task provided a static generalized view as to what Senior Leaders found important. In reality, Cyber SA priorities are very liquid, changing with context, mission focus, and current priorities. Additionally, the static nature of this exercise does not account for the frequency that a Senior Leader may need to see information. For example, Global Security Alert Levels may be considered to be of high importance, however, they change on a weekly basis, so a Senior Leader may only require to see it once a week. Future card sorts should add both context and time. Rather than focusing on overall, or general prioritizations, participants should be engaged with mission context, and issues of temporality, and to see how the sorts change. These exercises, and the interaction between them will help provide further context and understanding to their changing Cyber SA needs.

References

1. Endsley, M.R., Rodgers, M.D.: Situation awareness information requirements analysis for en route air traffic control. In: Proceedings of the Human Factors and Ergonomics Society Annual Meeting, pp. 71–75. SAGE Publications, Los Angeles (1994)
2. Kaber, D.B., Perry, C.M., Segall, N., McClernon, C.K., Prinzel III, L.J.: Situation awareness implications of adaptive automation for information processing in an air traffic control-related task. Int. J. Ind. Ergonom. **36**(5), 447–462 (2006)
3. Rodgers, M.: Human Factors Impacts in Air Traffic Management. Routledge, London (2017)
4. Harrald, J., Jefferson, T.: Shared situational awareness in emergency management mitigation and response. In: HICSS 2007 40th Annual Hawaii International Conference on System Sciences, pp. 23–23. IEEE (2007)
5. McNeese, M.D., Connors, E.S., Jones, R.E., Terrell, I.S., Jefferson Jr, T., Brewer, I., Bains, P.: Encountering computer-supported cooperative work via the living lab: application to emergency crisis management. In: Proceedings of the 11th International Conference of Human-Computer Interaction (2005)
6. McNeese, M.D., Mancuso, V.F., McNeese, N.J., Endsley, T., Forster, P.: Using the living laboratory framework as a basis for understanding next-generation analyst work. Security, and Sensing, Baltimore, Maryland, Paper presented at the SPIE Defense (2013)
7. Endsley, M.R.: Toward a theory of situation awareness in dynamic systems. Hum. Factors **37**(1), 32–64 (1995)
8. Joint Publications, Joint Publications 3–12 Cyberspace Operations. http://www.jcs.mil/Doctrine/Joint-Doctrine-Pubs/3-0-Operations-Series/ (2018)
9. Barford, P., et al.: Cyber SA: Situational awareness for cyber defense. In: Cyber Situational Awareness, pp. 3–13. Springer, Boston (2010)
10. Endsley, M.R., Garland, D.: Theoretical underpinnings of situation awareness: a critical review. In: Endsley, M.R., Garland, D. (eds.) Situation Awareness Analysis and Measurement, pp. 3–32. Taylor & Francis, Mahwah (2000)
11. Asgharpour, F., Liu, D., Camp, L.J.: Mental models of security risks. In: International Conference on Financial Cryptography and Data Security, pp. 367–377. Springer, Heidelberg (2007)
12. Falks, A., Hyland, N.: Gaining user insight: a case study illustrating the card sort technique. Coll. Res. Libr. **61**(4), 349–357 (2000)
13. Fleiss, J.L.: Measuring nominal scale agreement among many raters. Psychol. Bull. **76**(5), 378–382 (1971)
14. Star, S.L., Griesemer, J.R.: Institutional ecology, translations' and boundary objects: amateurs and professionals in Berkeley's museum of vertebrate zoology, 1907-39. Soc. Stud. Sci. **19**(3), 387–420 (1989)

Over-the-Shoulder Attack Resistant Graphical Authentication Schemes Impact on Working Memory

Jeremiah D. Still(✉) and Ashley A. Cain

Psychology of Design Laboratory, Department of Psychology,
Old Dominion University, Norfolk, VA, USA
{jstill, acain001}@odu.edu

Abstract. Alphanumeric passwords are the most commonly employed authentication scheme. However, technical security requirements often make alphanumeric authentication difficult to use. Researchers have developed graphical authentication schemes to help strike a balance between security requirements and usability. However, replacing characters with pictures has introduced both negative (security vulnerabilities) and positive (memorability benefits) outcomes. We are aware of the noteworthy long-term memory advantages of graphical passcodes, but little is known about the impact on users' limited working memory resources. Authentication is always a secondary task, which probably consumes working memory. This pilot study examines the impact graphical authentication schemes (Convex-Hull Click; Use Your Illusion; What You See is Where you Enter) have on working memory (Verbal; Spatial; Central Executive). Our findings suggest that graphical authentication schemes impact on working memory varies. This work shows that further investigation is needed to understand the complex relationship between scheme design and working memory.

Keywords: Human factors · Cybersecurity · Authentication · Working memory · Human-Computer interaction · Usability

1 Introduction

Authentication protects valuable information by requiring the user to validate their identity. A user is granted access to a system if they can confirm something they know (traditional password), something they have (cryptographic key), or something they are (fingerprint). Authenticating with alphanumeric passwords is conventional [1, 2]. However, this classic scheme for authentication is becoming too easy for attackers to overcome. Users are noncompliant, which offers hackers with a soft target (e.g., do not use the entire password dimensional space, reuse passwords, share passwords, record passwords physically). A recent report shows that 85% of users create passwords with personal information, and 50% use the same password for multiple accounts [3]. Maintaining strong passwords requires users adhere to numerous technical requirements. These additional security requirements can make password authentication difficult [1]. For example, passwords have to be long and complex to help defend against

T. Ahram and W. Karwowski (Eds.): AHFE 2019, AISC 960, pp. 79–86, 2020.
https://doi.org/10.1007/978-3-030-20488-4_8

brute force attacks, but this also can make the password more difficult to remember. Any next-generation authentication scheme needs to strike a balance between security requirements and usability [4]. Popular graphical next-generation schemes involve the use of pictures instead of alphanumeric characters. People can quickly and easily remember pictures versus strings of letters, symbols, and numbers. However, this very strength of graphical authentication presents a weakness for security. A casual attacker can glance over-the-shoulder of a user and steal their credentials. A recent report suggests that 81% of users are mindful of over-the-shoulder attacks in public places [3]. The recognition of this security vulnerability has produced over a decade of possible deterrents against casual bystanders. They can be classified as schemes that group targets among distractors, translate them to another location, and disguise the appearance of targets. These schemes can help prevent Over-the-Shoulder Attacks (OSA) [6].

These OSA resistant graphical authentication schemes show impressive passcode retention. The previous literature over the last decade strongly suggests that visually rich stimuli such as pictures facilitate memorability [7]. Further, images offer extra sensory data (c.f., letter, numbers, and special characters), which helps encoding and later retrieval (i.e., picture superiority effect; [8]). According to Still, Cain & Schuster (2017), recognizing and selecting graphics is often easier than recalling and generating a password [9]. This is similar to the effort required to recognize a correct answer within a multiple-choice test compared with writing an essay. A multiple-choice question provides the tester with retrieval cues that facilitate long-term memory access [10]. According to Cain and Still (2018), participants were able to remember graphical passcodes following a three-week delay; however, almost no one remembered their alphanumeric passcode [5]. Even when users are provided only a few practice trials, they do not forget system assigned graphical passcodes weeks later.

Even with a significant memory advantage, there are other factors to consider that might harm ease of use. For instance, a complex login procedure could make authentication difficult. More specifically the amount/type of working memory an authentication system drains might also predict its perceived "ease of use." Currently, no one has examined the impact graphical authentication has on working memory. We find this surprising given authentication is a secondary task (i.e., users simply want past security to work towards a goal). Further, authentication is often required numerous times throughout the day. The previous literature has focused on making authentication schemes efficient [11] and recognizing standards [12]. However, the cost of authentication can go beyond simply inconveniencing users. It is likely that schemes that consume our limited working memory resources could harm our ability to remember primary task information (e.g., the reason you needed to authenticate in the first place).

Baddeley described working memory as a conscious workbench with a limited capacity. Using the workbench is effortful and directed, but flexible as we assemble, or encode, information into long-term memory. Working memory plays an important role in interface interactions. Baddeley's model of working memory is composed of three main components: central executive, verbal (visuo-spatial sketchpad), and visual (articulatory loop) [13]. The verbal storage is fleeting with a duration of approximately 2000 ms. However, rehearsal, like subvocalization, can maintain the memory from being lost. Classic examples include remembering items in a list such as drink preferences for an upcoming party. The spatial storage helps users resolve every visuospatial complexity from high to low-level cognitive operations, whether a user is finding their way home to avoid new local flooding or integrating compatible motor and haptic information. The central executive allocates limited cognitive resources and combines information between the spatial and verbal components (for review see [14]). Our experiment separately loads the three main components of working memory (executive, spatial, verbal), while participants authentication using the most popular OSA-resistant authentication schemes (Convex-Hull Click (CHC): [15], Use Your Illusion (UYI): [16], What You See is Where you Enter (WYSWYE): [17]). Our findings will reveal whether a graphical authentication scheme impacts working memory, which could divert resources from the user's primary task.

2 Method

2.1 Participants

Eighteen undergraduate volunteers participated for course credit. Our sample was composed of 14 females, 15 native English speakers, and they reported having an average of 7 h of daily computer use (*SD* = 3.05). Two participants had to be excluded due to data collection errors, which resulted in missing data or placement in the wrong condition. We needed participants who were able to demonstrate an ability to authenticate. Following removal of participants with a login proficiency below 80%, only nine participants remained; three per authentication type. The remaining 9 participants were able to successfully authenticate; (UYI: *M* = 99; *SD* = 1.73), (WYSWYE: *M* = 92; *SD* = 8.54), (CHC: *M* = 89; *SD* = 3.46). Clearly, our brief graphical authentication training procedure was probably not sufficient; (CHC: *M* = 73; *SD* = 6.81), (WYSWYE: *M* = 69; *SD* = 10.15).

2.2 Stimuli and Apparatus

We created three prototypes of graphical authentication schemes for this study. The schemes were based on Convex-Hull Click (CHC; [15]), Use Your Illusion (UYI; [16]), and What You See is Where You Enter (WYSWYE; [17]). We also created a working memory task. CHC, UYI, WYSWYE, and the working memory task were presented on a Windows desktop computer with a 24-inch monitor; the graphical passcodes' presentation and data collection was controlled using Paradigm©, and the working memory task data collection and presentation was controlled using E-Prime©.

CHC. Convex-Hull Click consisted of icons on a 10 × 15 grid. The icons came from an online, open source database (http://www.fatcow.com/free-icons). The grid was 4138 × 1126 pixels. Each icon was 55 × 45 pixels. The passcode consisted of three system-assigned icons. Target icons were never located in a straight line. Because there were three target icons, they would always form a triangle shape on the grid. A correct login occurred when a participant selected an icon inside of the convex-hull created by the three icons. They were instructed not to click directly on target icons and not to hover the mouse cursor over their target icons. The researcher provided verbal feedback of correct or incorrect after each authentication attempt. After each attempt, the icons were reconfigured. This prototype represents a collection of OSA-resistant scheme implementations found in the literature [18–26].

UYI. Use Your Illusion was presented as images in a 3 × 3 grid that were degraded by removing specific details but retaining general colors and shapes (for more information see Tiller et al. [27]. The grid was 774 × 571 pixels. Each image was 213 × 175 pixels. A passcode consisted of three system-assigned images. A correct login occurred when a participant selected the degraded versions of each of their three targets across three subsequent grids. The researcher provided verbal feedback of correct or incorrect after each authentication attempt. After each attempt, the images were reconfigured. Again, this prototype represents a larger collection of OSA-resistant schemes found in the literature [28–31, 33–38].

WYSWYE. The interface for What You See is Where You Enter showed a 5 × 5 grid of images on the right side of the screen. The grid of images was 715 × 549 pixels. Each image was 139 × 103 pixels. A blank 4 × 4 grid was on the left side. The blank grid was 578 by 459 pixels. The blank cells were 139 × 103 pixels. A passcode consisted of four system-assigned images. Participants had to perform mental operations before logging in. First, they had to mentally delete a row and column that did not contain a target. Then, they would mentally shift the remaining cells together. This would mentally reduce the 5 × 5 grid to a 4 × 4 grid. They logged in by clicking the locations of their four targets on the blank grid. The researcher provided verbal feedback of correct or incorrect after each authentication attempt. The images were reconfigured for every attempt. Like the others, this prototype represents a large collection of OSA-resistant schemes found in the literature [39–47].

Working Memory Task. During each authentication trial, working memory was loaded with verbal, spatial, or central executive information. Participants were asked to hold the heavy load in working memory while they logged in. The working memory load was reported following authentication to verify the information was actually being maintained. All three components of Baddeley's working memory model were examined (c.f., [48]). Verbal working memory was loaded by presenting four letters. Then the letters disappeared, and the participants were asked to type the four letters after logging in. Spatial working memory was loaded by presenting four dots on a 3 × 3 grid. Then the dots disappeared and were replaced by a 3 × 3 grid of numbers. The participants were asked to type the four letters that were in the locations where the dots had been. Central executive working memory was loaded by presenting four letters on a 3 × 3 grid. Then the letters disappeared and were replaced by a 3 × 3 grid of numbers. The participants were asked to type the numbers that were in the locations where the letters had been and were asked to type the letters.

2.3 Procedure

Participants were run individually. They were seated in front of two desktop computers. On one computer Paradigm© presented the graphical schemes, and on the other E-Prime© presented the working memory task. Each participant interacted with a single graphical scheme. However, every participant received all the different types of working memory load. The working memory type presentation order was not completely counterbalanced. Before each authentication trial, the working memory task presented items to be remembered, and after each trial the participants typed the items to the best of their ability. Participants were provided with instructions, 20 authentication scheme practice trials, and three working memory task practice trials. The experimenter provide feedback on whether participants had correctly authenticated or not on every trial.

3 Results

All the statistical tests used an alpha level of .05. Clearly, the overall numerical trends show the negative impact authentication has on information held in working memory. To explore our manipulations, a repeated measures ANOVA examining working memory type (verbal, spatial, central executive) as a function of authentication scheme (between-subject factor: UYI, CHC, WYSWYE) revealed significant main effects on working memory accuracy. The analysis revealed a within-subjects main effect of Working Memory Type, $F(2,12) = 16.84$, $p < .001$, $n_p^2 = .70$. And, the between-subjects main effect of Authentication Scheme, $F(2,6) = 9.71$, $p = .013$, $n_p^2 = .764$. However, the interaction between Working Memory Type X Authentication Scheme was not found to be significant, $F(4,12) = .601$, $p = .669$, $n_p^2 = .05$. Post-hoc tests were conducted using Bonferroni with adjusted alpha levels to avoid Type I error inflation across multiple comparisons.

Working Memory Type. Results indicated that Spatial working memory ($M = .79$, $SEM = .04$) was best preserved compared to Verbal ($M = .54$, $SEM = .06$) and Central Executive ($M = .34$, $SEM = .07$), $p < .05$. However, Verbal working memory and Central Executive were not significant different from each other, $p = .139$.

Authentication Scheme. It was revealed that the CHC and WYSWYE authentication schemes were the only schemes to be statistically different. Specifically, the CHC ($M = .67$, $SEM = .04$) had less negative impact on working memory accuracy compared with the WYSWYE scheme ($M = .44$, $SEM = .08$), $p = .014$, $d = 1.466$ (see Fig. 1).

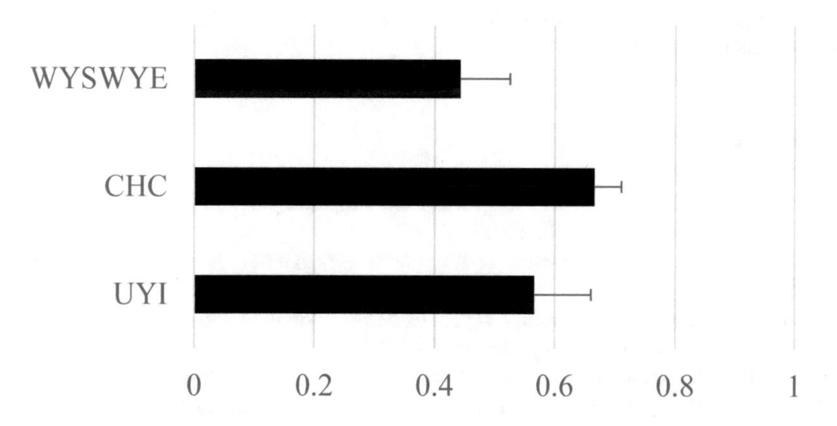

Fig. 1. Proportion of correctly reported working memory trials by authentication scheme. The error bars represent standard error of the mean.

4 General Discussion

Authentication schemes must afford us with easy access to our data, while maintaining their security. This is a difficult balancing act between user and technical requirements. The conventional alphanumeric scheme has quickly evolved into a soft target. Attackers know passwords are being reused and contain personal information. In an attempt to strengthen passwords, the technical requirements for a strong password are increased. This makes remembering a password difficult. Therefore, authentication developers have been searching for a next-generation solution. One popular solution proposed in the literature is graphical authentication.

These schemes offer users with superior memorability compared to password-based schemes. However, a graphical authentication system usually requires more complex interactions than simply typing in a password. For example, a user might be asked to search for an icon among many decoys. It is recognized logging in with a graphical authentication scheme is less efficient compared with conventional schemes (e.g., password or PIN). This could annoy users given they need to login numerous times throughout the day. However, harming working memory can go beyond subjective dissatisfaction. Surprisingly, no one has examined the impact of graphical authentication on the working memory system. Stealing limited working memory resources from the primary task to authenticate might harm our primary task performance.

This pilot study found that working memory performance varies by graphical authentication scheme. Clearly, authentication interactions influence the user's ability to remember information being held in working memory for the primary task. We believe that a larger sample size will reveal additional significant differences between authentication schemes given the medium to large effect sizes (i.e., range = .66–.81). Future work needs to explore the interaction between working memory types and authentication schemes. We suggest a follow-up study, increases the number of participants in the study and introduces a low-load and password conditions. This will allow researchers to explore the actual impact of working memory type independent of task difficulty.

Acknowledgments. We thank Paige Duplantis, Lauren Tiller, and Ayobami Fakulujo for their assistance collecting data.

References

1. Grawemeyer, B., Johnson, H.: Using and managing multiple passwords: a week to a view. Interact. Comput. **23**, 256–267 (2011)
2. Zviran, M., Haga, W.J.: Password security: an empirical study. J. Man. Info. Sys. **15**, 161–185 (1999)
3. Cain, A.A., Edwards, M.E., Still, J.D.: An exploratory study of cyber hygiene behaviors and knowledge. J. Info. Secur. App. **42**, 36–45 (2018)
4. Still, J.D.: Cybersecurity needs you! ACM Interact. (May + June: Feature). **23**, 54–58 (2016)
5. Cain, A.A., Still, J.D.: Usability comparison of over-the-shoulder attack resistant authentication schemes. J. Usab. Stud. **13**, 196–219 (2018)

6. Cain, A.A., Werner, S., Still, J.D.: Graphical authentication resistance to over-the-shoulder-attacks. In: Proceedings CHI Conference Extended Abstracts, pp. 2416–2422 (2017)
7. Biddle, R., Chiasson, S., Van Oorschot, P.C.: Graphical passwords: learning from the first twelve years. ACM Comp. Sur. (CSUR) **44**, 1–25 (2012)
8. Mintzer, M.Z., Snodgrass, J.G.: The picture superiority effect: support for the distinctiveness model. Amer. J. Psyc. **112**, 113–146 (1999)
9. Still, J.D., Cain, A., Schuster, D.: Human-centered authentication guidelines. Info. Comp. Sec. **25**, 437–453 (2017)
10. Tulving, E., Thomson, D.M.: Encoding specificity and retrieval processes in episodic memory. Psyc. Rev. **80**, 352–373 (1973)
11. Werner, S., Hauck, C., Masingale, M.: Password entry times for recognition-based graphical passwords. Proc. Hum. Factors Ergon. Soc. Annu. Meet. **60**, 755–759 (2016)
12. Braz, C., Robert, J.: Security and usability: the case of the user authentication methods. In: Proceedings of the 18th International Conference on Association Francophone d'Interaction Homme-Machine, 199–203 (2006)
13. Baddeley, A.: Working memory. Science **255**, 556–559 (1992)
14. Logie, R.H.: Retiring the central executive. Q. J. Exp. Psychol. (2016). advance online publication
15. Wiedenbeck, S., Waters, J., Sobrado, L., Birget, J. C.: Design and evaluation of a shoulder-surfing resistant graphical password scheme. In: Proceedings of the Working Conference on Advanced Visual Interfaces, pp. 177–184 (2006)
16. Hayashi, E., Dhamija, R., Christin, N., Perrig, A.: Use your illusion: secure authentication usable anywhere. In: Proceedings of the 4th Symposium on Usable Privacy and Security, pp. 35–45 (2008)
17. Khot, R.A., Kumaraguru, P., Srinathan, K.: WYSWYE: shoulder surfing defense for recognition based graphical passwords. In: Proceedings of the 24th Australian CHI Conference, pp. 285–294 (2012)
18. Ankush, D.A., Husain, S.S.: Authentication scheme for shoulder surfing using graphical and pair based scheme. Intern. J. Adv. Res. Comp. Sci. Mang. Stud. **2**, 161–166 (2014)
19. Behl, U., Bhat, D., Ubhaykar, N., Godbole, V., Kulkarni, S.: Multi-level scalable textual-graphical password authentication scheme for web based applications. J. Electron. Commun. **3**, 166–124 (2014)
20. Chen, Y.L., Ku, W.C., Yeh, Y.C., Liao, D.M.: A simple text-based shoulder surfing resistant graphical password scheme. In: IEEE ISNE, pp. 161–164 (2013)
21. Joshuva, M., Rani, T.S., John, M.S.: Implementing CHC to counter shoulder surfing attack in PassPoint–style graphical passwords. Intern. J. Adv. Net. App. **2**, 906–910 (2011)
22. Kiran, T.S.R., Rao, K.S., Rao, M.K.: A novel graphical password scheme resistant to peeping attack. Int. J. Comput. Sci. Inf. Technol. **3**, 5051–5054 (2012)
23. Manjunath, G., Satheesh, K., Saranyadevi, C., Nithya, M.: Text-based shoulder surfing resistant graphical password scheme. Intern. J. Comp. Sci. Info. Tech. **5**, 2277–2280 (2014)
24. Rao, K., Yalamanchili, S.: Novel shoulder-surfing resistant authentication schemes using text-graphical passwords. Int. J. Inf. Secur. **1**, 163–170 (2012)
25. Vachaspati, P.S.V., Chakravarthy, A.S.N., Avadhani, P.S.: A novel soft computing authentication scheme for textual and graphical passwords. Intern. J. Comp. App. **71**, 42–54 (2013)
26. Zhao, H., Li, X.: S3PAS: a scalable shoulder-surfing resistant textual-graphical password authentication scheme. In: 21st AINAW, vol. 2, pp. 467–472 (2007)
27. Tiller, L., Cain, A., Potter, L., Still, J.D.: Graphical authentication schemes: balancing amount of image distortion. In: Ahram, T., Nicholson, D. (eds.) Advances in Human Factors in Cybersecurity, pp. 88–98 (2019)

28. Cain, A.A., Still, J.D.: A rapid serial visual presentation method for graphical authentication. In: Nicholson, D. (ed.) Advances in Human Factors Cybersecurity, pp. 3–11. Springer, Cham (2016). https://doi.org/10.1007/978-3-319-41932-9_1
29. Gao, H., Guo, X., Chen, X., Wang, L., Liu, X.: Yagp: yet another graphical password strategy. In: Computer Security Applications Conference, pp. 121–129 (2008)
30. Ghori, F., Abbasi, K.: Secure user authentication using graphical passwords. J. Ind. Stud. Res. **11**, 34–40 (2013)
31. Hui, L.T., Bashier, H.K., Hoe, L.S., Kwee, W.K., Sayeed, M.S.: A hybrid graphical password scheme for high-end system. Aust. J. Bas. App. Sci. **8**, 23–29 (2014)
32. Jenkins, R., McLachlan, J.L., Renaud, K.: Facelock: familiarity-based graphical authentication. Peer J. **2**, 1–24 (2014)
33. Lin, D., Dunphy, P., Olivier, P., Yan, J.: Graphical passwords & qualitative spatial relations. In: Proceedings of Symposium on Usable Privacy and Security, pp. 161–162 (2007)
34. Meng, Y., Li, W.: Enhancing click-draw based graphical passwords using multi-touch on mobile phones. In: IFIP Conference, pp. 55–68 (2013)
35. Nicholson, J.: Design of a Multi-touch shoulder surfing resilient graphical password. B. Sci. Info. Sys. (2009)
36. Sasamoto, H., Christin, N., Hayashi, E.: Undercover: authentication usable in front of prying eyes. In: Proceedings of the SIGCHI Conference, pp. 183–192 (2008)
37. Yakovlev, V.A., Arkhipov, V.V.: User authentication based on the chess graphical password scheme resistant to shoulder surfing. Auto. Con. Comp. Sci. **49**, 803–812 (2015)
38. Zakaria, N.H., Griffiths, D., Brostoff, S., Yan, J.: Shoulder surfing defense for recall-based graphical passwords. In: Proceedings of Seventh Symposium on Usable Privacy and Security, pp. 6–18 (2011)
39. Bianchi, A., Oakley, I., Kim, H.: PassBYOP: bring your own picture for securing graphical passwords. IEEE Trans. Hum. Mach. Syst. **46**, 380–389 (2016)
40. Brostoff, S., Inglesant, P., Sasse, M.A.: Evaluating the usability and security of a graphical one-time PIN system. In: Proceedings of the 24th BCS Interaction Specialist Conference, pp. 88–97 (2010)
41. De Luca, A., Hertzschuch, K., Hussmann, H.: ColorPIN: securing PIN entry through indirect input. In: Proceedings of the SIGCHI, pp. 1103–1106 (2010)
42. Gao, H., Liu, X., Dai, R., Wang, S., Chang, X.: Analysis and evaluation of the colorlogin graphical password scheme. In: Fifth International Conference on Image and Graphics, pp. 722–727 (2009)
43. Gupta, S., Sahni, S., Sabbu, P., Varma, S., Gangashetty, S.V.: Passblot: a highly scalable graphical one time password system. Intern. J. Net. Sec. App. **4**, 201–216 (2012)
44. Kawagoe, K., Sakaguchi, S., Sakon, Y., Huang, H.H.: Tag association based graphical password using image feature matching. In: International Conference on Database Systems for Advanced Applications, pp. 282–286 (2012)
45. Lashkari, A.H., Manaf, A.A., Masrom, M.: A secure recognition based graphical password by watermarking. In: 11th International Conference on Computer and Information Technology, pp. 164–170 (2011)
46. Perkovic, T., Cagalj, M., Rakic, N.: SSSL: shoulder surfing safe login. In: 17th International Conference Software, Telecommunications & Computer Network, pp. 270–275 (2009)
47. Zangooei, T., Mansoori, M., Welch, I.: A hybrid recognition and recall based approach in graphical passwords. In: Proceedings of the 24th Australian CHI Conference, pp. 665–673 (2012)
48. Still, J.D., Dark, V.J.: Examining working memory load and congruency effects on affordances and conventions. Int. J. Hum Comput Stud. **68**, 561–571 (2010)

Comparative Evaluation of Security and Convenience Trade-Offs in Password Generation Aiding Systems

Michael Stainbrook and Nicholas Caporusso(✉)

Fort Hays State University, 600 Park Street, Hays 67601, USA
mjstainbrook@mail.fhsu.edu, n_caporusso@fhsu.edu

Abstract. A strong password is considered the most important feature for the security of any account credentials. In the last decades, several organizations focused on improving its strength and produced awareness initiatives and security guidelines on how to create and maintain secure passwords. However, studies found that users perceive security and convenience as a trade-off, and they often compromise password strength in favor of a key phrase that is easier to remember and type. Therefore, nowadays websites and applications implement password generation aiding systems (PGAS) that help, and even force, users to create more secure passwords. Nowadays, several types of PGAS are available, each implementing a different strategy for stimulating users in crating stronger and more secure passwords. In this paper, we present the results of a study in which we compared six different PGAS and evaluated their performance in terms of security and convenience, with the aim of suggesting the system that has the most beneficial trade-off depending on the type of application.

Keywords: Password meters · Cybersecurity · Credentials

1 Introduction

In the last decade, novel access systems implementing sophisticated credentials, such as, two-factor authentication and biometric identification [1], have increasingly been utilized for protecting the security and privacy of information in devices and accounts. Nevertheless, text-based passwords are still the most common authentication method for accessing websites, electronic mailboxes, and other types of accounts (e.g., wireless networks). However, the main limitation of passwords lies in the paradox of the trade-off between security and convenience: strong and secure passwords typically are inconvenient and difficult to remember [2]. As a result, research showed that users tend to utilize a total of 3–6 passwords shared between multiple accounts, even if they adopt a password manager; moreover, they prefer to create key phrases that are easy to remember by including information that is meaningful to them (such as, important dates and names) which, in turn, affects their security [3].

Password generation aiding systems (PGAS) have been developed to address the issue and help users increase the strength of their credentials by enforcing specific requirements (e.g., minimum length, presence of special symbols, or entropy) when

T. Ahram and W. Karwowski (Eds.): AHFE 2019, AISC 960, pp. 87–96, 2020.
https://doi.org/10.1007/978-3-030-20488-4_9

they create a new password. Although they provide users with feedback about security score, research showed that users typically approach most PGAS as a checklist, and they do not result in any security improvement beyond the lowest mandatory level [3]. Indeed, there are several types of PGAS and they have very different strategy, design, and interface characteristics. However, only a few studies took into consideration the performance implications of the different components of user experience in password generation aiding systems, such as, interface, usability, and type and timeliness of feedback.

2 Related Work

Text-based passwords can be considered an imperfect method of authentication. For users to have strong and secure key phrases, they must create them in ways that make them inconvenient and difficult to remember [2]. Previous research studied users' behavior in password creation: users tend to reiterate the same key phrase over multiple accounts and prefer to use alphanumeric strings that are easy to remember and type whilst including something meaningful to them [3]. This challenge of generating passwords falls into the trade-off of security and convenience, where a convenient and easy to remember password is inconvenient for the user and difficult to remember [4]. Developers have attempted to influence users to create more secure passwords through the implementation of real-time feedback systems. These systems generally attempt to give users a real-time security evaluation of their password, considering the entropy, length, characters used, and looking for non-sequential letters and numbers. The calculated score of the password meter gives the user an estimate into the security of their password.

Research has attempted to determine the effectiveness of password feedback systems in aiding users to create more secure passwords. Studies have found that in all cases password feedback systems have influenced users with low password scores to create a more secure password [5]. The feedback systems studied have varied from simplistic password meters, to detailed systems and peer feedback. When users were questioned about which feedback system provided the best information and would most likely use, respondents selected a password meter which provided the most information about their password; however, the users also responded that this detailed password feedback was the most difficult to understand as well [6]. Another study of password feedback focused on a peer feedback password meter. This meter would show how the newly created password compared to other user's passwords on the site. Researchers found that the peer feedback system did not provide a significant effect on motivating users to increase the security of their password unless explicit instructions were included with the meter [6].

Password feedback mechanisms may help to improve users with low password security; however, it seems that context and account type may be a key identifying factors for if the feedback is taken into consideration by the user. A study found that password feedback systems did not increase password security for users creating unimportant accounts, even though they are commonly deployed on such sites. However, for accounts that contain sensitive information users appeared to take the password feedback into account when changing their passwords on these sites. Additionally, researchers found that providing a password meter for users when changing passwords helps to influence users to create stronger passwords, opposed to showing a password meter when first creating the account [7].

Password meters are currently active in many of Alexa's 100 most visited global sites. Out of the top 100 sites, 96 allow the ability for users to create a password and account, out of the 96, 73% gave some sort of feedback to users on their passwords, many of the sites used similar or the same password meter ranging from a bar meter to checkmarks systems after meeting requirements [8]. These systems were dynamically updated as the user types and considered the length, characters used, and occasionally blacklisted words [10]. A large-scale study on the effectiveness of password meters found that overall, password meters do change user behavior when interacting with them. Researchers found that users seeing a password meter nudged users to create longer passwords. Furthermore, findings suggested that the visual component of a password meter did not lead to significant differences. Users presented with a dancing bunny password meter reacted the same as those presented with traditional password feedback; however, the combination of text and a visual component was an important factor in the effectiveness. Additionally, the researchers found that users substantially changed their behavior when they were presented with stringent meters that would purposefully lower the security score of the password to motivate users to create a more secure password. Moreover, the stringent meters did motivate users to change their passwords; however, users also reported these meters to be annoying and frustrating to use [9].

Most research agrees that password feedback can help improve user's passwords [6, 7, 9]; however, password meters with requirements may be viewed by some users as a checklist. A study on the impact of feedback on password-creation behavior found that some users would quit adding characters and length to their passwords after fulfilling the password feedback requirements, such as: minimum of eight characters, at least one upper and lower-case letter, etc. Researchers theorized that some users may view the feedback as requirements and quit improving their password security after fulfilling them as they give the user a feeling of finality from reaching the requirements. Whereas, in situations where the password feedback did not have requirements, users may not be aware that the requirements were met and add additional security such as length and special characters. Another, theory the researchers had about why users may stop adding security after meeting requirements is that they rely on the feedback system for security. Therefore, users trust the feedback system, relying on it to help them create a secure password. Ultimately, the researchers recommend password feedback systems prompt users to continue adding security mechanisms to their passwords after fulfilling password requirements [10].

Implementing password feedback systems and meters may help to improve user passwords by giving explicit instructions and providing a visual representation, usually in a horizontal bar that increases and fills as the security increases. Moreover, this does not help improve the problem of users and poor password management. A large-scale study of user-password interaction found that the average user had 6.5 passwords shared between 3.9 sites [8]; another study found that 63.8% users reported using their password elsewhere, despite being aware of security practices [7]. This illustrates the problem of getting users to take real-time password feedback mechanisms and instructions into account when creating a new account. Moreover, it seems that younger generations may have the worst password security practices. A study found that younger users tend to ignore password feedback requirements for creating secure passwords: they may persuade themselves that the contents of their accounts are of little use to malicious users, not taking into consideration that their login credentials for secure accounts may be reused or similar [9]. In addition to password security, other studies [10] analyzed current issues and practices in enforcing username security. In conclusion, given current trends in cybercrime [12] and the rapidly changing dynamics of Internet consumption (e.g., cybermigration [13]), PGAS keep playing a fundamental role in creating awareness and fostering security.

3 Strategies in Password Generation Aiding Systems

In this Section, we review the most common strategies utilized in PGAS for helping users increase the security of their passwords: reactive techniques include suggesting guidelines, enforcing requirements, and giving feedback, whereas proactive methods automatically generate a secure key phrase for the user. They are detailed in Fig. 1. Although their functionality might vary depending on the type of information in an account, their general purpose is to improve the trade-off between security, that is, effort to crack, and convenience (i.e., effort to generate and use).

3.1 Showing Password Guidelines

In its simplest form, a password generation aiding strategy would display the guidelines for creating a more secure key phrase. As discussed earlier, they might vary depending on the type of information that is maintained in the digital resource. Moreover, in its simplest form, a PGAS would list password criteria as a suggestion to improve the security of the key phrase, such as, minimum length, use of a mix of numbers and letters with mixed case, and adding some special symbols. However, it would not enforce them, enabling users to proceed with account registration even if the password does not meet the requirements. As a result, users are left with the responsibility of creating a strong password, and they can opt for a key phrase that they consider convenient, even if less secure.

3.2 Enforcing Password Requirements

The majority of PGAS implemented in today's systems fall in this category: in addition to showing the requirements, they enforce them by preventing users from creating an account or changing their passwords if the chosen key phrase does not meet the specified criteria. The main difference with the system described in 3.1 is in that systems enforce specifications as requirements. As a result, the user must generate a password that meets the level of security of the system. Furthermore, this type of PGAS typically provide feedback, either in real-time or after form submission, about the items that have not been addressed yet. The advantage of this system is two-fold: (1) it prevents creating passwords that are below a specific strength and (2) it educates users about security by having them practice.

3.3 Strength Meter

As the level of security of a password is directly proportional to its complexity, it can be increased by manipulating two dimensions, that is, length and content diversity. Primarily, key phrase guidelines and requirements have the purpose of supporting users in generating a key phrase that is compliant with both criteria. Nevertheless, a secure password could be achieved by manipulating either length or content. As a result, instead of specifying requirements, strength meters evaluate password entropy without taking into consideration its specific components. Unfortunately, visualization in the form of labels might create some ambiguity because password meters might yield different results depending on the security requirements of the system. As a result, a password that is considered good by the PGAS of a website could be graded differently in another that has higher standards.

3.4 Time to Crack

Strength meters are a user-friendly abstraction of the concept of entropy: labels identifying password's strength are a good compromise between the need of protecting an account with a secure key phrase and the possibility of giving users the possibility of manipulating either of the two degrees of freedom (i.e., length or entropy). Moreover, they provide quick and convenient feedback. Similarly, systems using time-to-crack as a measure of password security display entropy as an estimate of the time that a brute-force attack would require to guess the password. Unfortunately, this type of PGAS suffer from the same limitations as strength meters. Moreover, time to crack depends on a variety of factors: its measurement might not be accurate and might significantly vary depending on the type of technique and resources used in an attack. Also, as cybercrime evolves, novel and sophisticated techniques are expected to be faster in breaching systems.

3.5 Comparing with User Base

As discussed in Sect. 2, studies showed that users tend to underestimate their own risk of being hacked. Thus, they tend to prefer convenience in favor of security though they are aware of cybersecurity practices. To this end, password requirements, strength meters, and time-to-crack estimates provide an impersonal measure of a hypothetical risk. Conversely, peer feedback has been studied as a strategy to incentivize individuals to adopt stronger key phrases by comparing them to others. As a result, this type of password meter engages users in a competitive dynamic aimed at fostering the creation of a key phrase that is stronger than the average password utilized by the members of a community. Indeed, this psychological trigger could be implemented by merely changing the interface of a standard password meter based on entropy.

3.6 Automatic Password Assignment

Studies about passwords demonstrated that users are very prone to forgetting their passwords when it is particularly complex, and thus, secure. Typically, this happens when users do not utilize password managers, or when the password is not stored in an account manager. Moreover, considering that nowadays users have dozens of accounts, they might have to generate, store or remember, and use many different passwords for websites and digital resources that they might access very seldom. As a result, several even started questioning the need of generating a password that is easy to remember: depending on the frequency of use of an account, users might want to create a password that is very secure and hard to remember, and then, either use a password manager or rely on the procedure for recovering their authentication information. Consequently, software and websites can automatically generate a secure key phrase for the user to store in a password manager, so that they can make sure that the security requirements. PGAS using automatic password assignment increase convenience of creating account credentials but rely on a third-party software or procedure (e.g., password manager or password recovery process) to cope with higher probability of password forgetfulness.

4 Experimental Study

In this Section, we detail a study in which we compared the different strategies and types of PGAS. Specifically, we evaluated the impact of human factors on the relationship between usability and performance in terms of security and convenience trade-off. To this end, we designed a web-based account creation system that implements the six different PGAS discussed in Sect. 3. We recruited a total of 115 subjects (33 females and 82 males aged 28 ± 9). Participants were sampled from a population having low to medium familiarity with IT (based on degree, background, time spent working with a desktop computer, and number of accounts) to prevent results from being biased by computer literacy.

In the experimental task, participants were directed to an Internet URL in which: (1) they were presented with a sign-up form incorporating a password generation aiding system selected at random and they were asked to create an account using a key phrase they never utilized in the past and save the information using the method of their choice; (2) they were redirected to an account activation page that had the purpose of interrupting their task; (3) they were presented with a sign-in form and they were asked to use their credentials. Finally, they were asked to fill a questionnaire in which they evaluated the convenience and usability of PGAS in the sign-up phase and in signing in. Participants were asked to realize the task six times (one for each type of system). The order of PGAS was selected at random to prevent training effect. The experimental software calculated the entropy of the password generated using each PGAS. Moreover, it acquired the time required for typing the key phrase at sign up (calculated from the first character entered until the user left the password field) and for signing in (calculated from page load until the user clicked the sign in button). The security criteria were the same for all systems: minimum 8 characters long, at least a number and an uppercase and a lowercase letter.

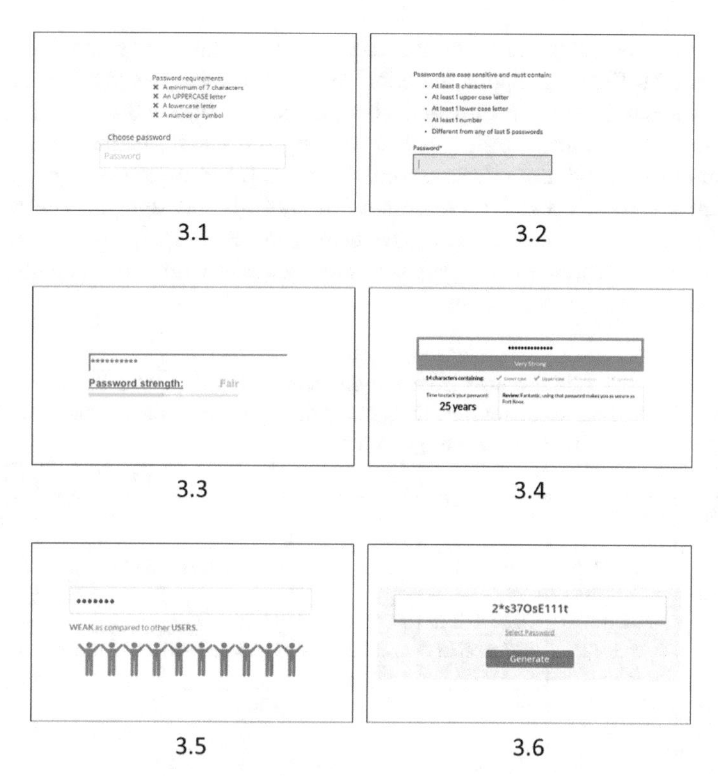

Fig. 1. Different types of Password Generation Aiding Systems and strategies as described in Sect. 3 and utilized in the study: showing password requirements (3.1) is the most basic approach to security, which can be enforced using systems that require the specifications to be met (3.2). Strength meters (3.3) score password robustness whereas systems based on time to crack (3.4) measure the security of a password in terms of time required to guess it, which can be estimated using length and entropy measurement. Also, non-conventional meters display the strength of the chosen password as compared to other users (3.5). Finally, a few software and websites are automatically generating secure passwords for the user when they register (3.6).

5 Results and Discussion

PGAS that suggest or require specific symbols resulted in the lowest overall score, as they required 4.20 s on average to create and use, had the least entropy and length, and were preferred last in terms of convenience. In line with findings in the literature, subjects utilized them as a checklist. Conversely, password meters had better results, though label-based systems had different outcome than meters based on time-to crack and peer comparison. Specifically, the former resulted lasts in terms of users' perception even if they had better performance, both as calculated effort to create and use, and as resulting entropy and length. Finally, PGAS that automatically generate a password resulted in the lowest effort to create and use and in the best preference. Although they require users to store the key phrase in password manager, they also prevent reusing the same word across multiple accounts and, thus, might have an impact on the overall security of the user. Table 1 shows a summary of all the experiment data. Figure 2 reports the effort to create and use the password, calculated as seconds spent in the sign up and sign in phases. Figure 3 indicates password strength, measured in terms of key phrase entropy and length. Perceived convenience in account creation and access was recorded using a Likert scale (see Fig. 4). Overall, users spent between 3 and 5 s on average to sign up and sign in. The entropy of password ranged from weak to strong depending on the PGAS, whereas the average length was 10 characters. Finally, systems were perceived similarly, though there is a statistical difference between them.

Table 1. Experiment data acquired from the different PAGS described in Sect. 3: password guidelines (3.1), password requirements (3.2), strength meter (3.3), time to crack indicator (3.4), peer strength meter (3.5), and password generator (3.6).

PGAS	3.1	3.2	3.3	3.4	3.5	3.6
Time to sign up (seconds)	4.15 ± 0.85	4.24 ± 0.88	4.11 ± 0.88	4.36 ± 1.44	4.12 ± 1.42	0.25 ± 0.83
Time to sign in (seconds)	4.42 ± 1.44	4.26 ± 1.33	3.78 ± 1.17	5.08 ± 1.68	4.91 ± 1.74	5.75 ± 3.06
Entropy (bits)	39.62 ± 7.30	41.22 ± 8.13	47.26 ± 8.80	43.28 ± 9.17	49.24 ± 8.30	47.5 ± 8.71
Length (characters)	9.45 ± 1.05	9.47 ± 1.15	11.25 ± 2.01	11.15 ± 1.92	11.12 ± 2.03	9.11 ± 0.57
Convenience in sign up (Likert)	3.38	3.47	3.41	4.36	4.29	4.85
Convenience in sign in (Likert)	3.12	3.21	3.77	2.77	3.04	2.74

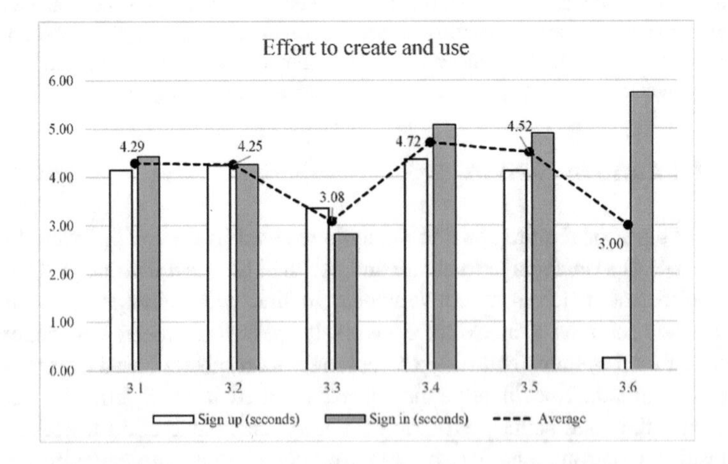

Fig. 2. Effort to create and use a password, measured in seconds required to generate it and to use it. PGAS that automatically generate a password (3.6) ranked best, though there was is a 5 s difference between sign up and sign in phases. Standard label-based password meters are the most efficient alternative.

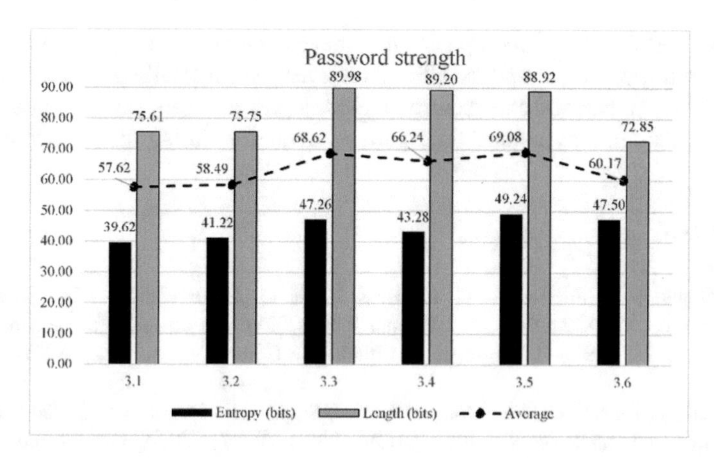

Fig. 3. Password strength calculated as entropy and length. The latter, which was initially measured in characters, was converted to bits (8 bits per character) for visualization purposes. Password meters resulted in the highest score of entropy and length. However, this was because only a few subjects changed the automatically-generated password in (3.6), which is statistically comparable to 3.3 in terms of entropy.

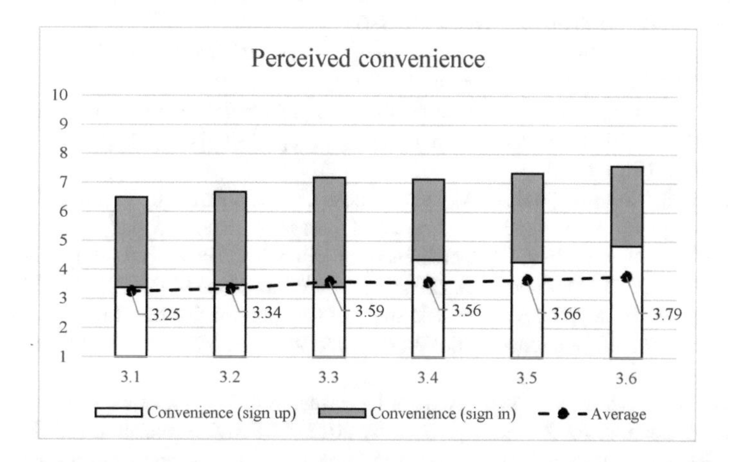

Fig. 4. Perceived convenience in the sign up and sign in phases, scored on a Likert scale. Systems that automatically generate a password for the user (3.6) ranked best, though individual steps differ significantly. Also, respondents indicated a preference for password meters over system that suggest or enforce password requirements.

6 Conclusion

In this paper, we presented an overview of the main types of currently available PGAS, we highlighted their differences, and we discussed the results of a comparative evaluation of their security performance and perceived usability. From our findings, we can conclude that interface design strategies lead to better trade-off between password robustness and convenience. Specifically, meters based on simple labels result in improved entropy, length, and usability. However, automatically-generated passwords score best in convenience and security, and they prevent users from reusing the same key phrase, though they require third-party systems for storing the password.

References

1. Bevilacqua, V., Cariello, L., Columbo, D., Daleno, D., Fabiano, M.D., Giannini, M., Mas-tronardi, G., Castellano, M.: Retinal fundus biometric analysis for personal identifications. In: International Conference on Intelligent Computing, pp. 1229–1237, September 2008
2. Bonneau, J., Herley, C., Van Oorschoto, P.C., Stajano, F.: Passwords and the evolution of imperfect authentication. Commun. ACM **58**(7), 78–87 (2015). https://doi.org/10.1145/2699390
3. Stainbrook, M., Caporusso, N.: Convenience or strength? Aiding optimal strategies in password generation. In: Proceedings of Advances in Human Factors in Cybersecurity, AHFE 2018. Advances in Intelligent Systems and Computing, vol 782 (2018). https://doi.org/10.1007/978-3-319-94782-2_3
4. Tam, L., Glassman, M., Vandenwauver, M.: The psychology of password management: a trade-off between security and convenience. Behav. Inf. Technol. **29**(3), 233–244 (2010). https://doi.org/10.1080/01449290903121386
5. Ciampa, M.: A comparison of password feedback mechanisms and their impact on password entropy. Inf. Manag. Comput. Secur. **21** (2013)
6. Dupuis, M., Khan, F.: Effects of peer feedback on password strength. In: APWG Symposium on Electronic Crime Research, San Diego, CA, pp. 1–9 (2018). https://doi.org/10.1109/ECRIME.2018.8376210
7. Egelman, S., Sotirakopoulos, A., Muslukhov, I., Beznosov, K., Herley, C.: Does my password go up to eleven? The Impact of password meter on password selection. In: Proceedings of the SIGCHI Conference on Human Factors in Computing Systems, pp. 2379–2388. ACM, New York (2013). https://doi.org/10.1145/2470654.2481329
8. Florêncio, D., Herley, C.: A large-scale study of web password habits. In: Proceedings of the 16th International Conference on the World Wide Web, pp. 657–666 ACM Press, New York (2007)
9. Ur, B., et al.: How does your password measure up? The effect of strength meters on password creation. In: Proceedings Security 2012, USENIX Association (2012)
10. Shay, R., et al.: A spoonful of sugar? The impact of guidance and feedback on password-creation behavior. In: Proceedings of the 33rd Annual ACM Conference on Human Factors in Computing Systems, pp. 2903–2912, April 2015
11. Caporusso, N., Chea, S., Abukhaled, R.: A game-theoretical model of ransomware. In: International Conference on Applied Human Factors and Ergonomics, pp. 69–78. Springer, Cham, July 2018. https://doi.org/10.1007/978-3-319-94782-2_7
12. Xiao, X., Caporusso, N.: Comparative evaluation of cyber migration factors in the current social media landscape. In: 2018 6th International Conference on Future Internet of Things and Cloud Workshops (FiCloudW), pp. 102–107. IEEE, August 2018. https://doi.org/10.1109/W-FiCloud.2018.00022
13. Fandakly, T., Caporusso, N.: Beyond passwords: enforcing username security as the first line of defense. In: International Conference on Applied Human Factors and Ergonomics. Springer, July 2019 (to be published)

Perceiving Behavior of Cyber Malware with Human-Machine Teaming

Yang Cai[1(✉)], Jose A. Morales[2(✉)], William Casey[2(✉)], Neta Ezer[3(✉)], and Sihan Wang[1]

[1] Cylab, Carnegie Mellon University, 4700 Forbes Ave, Pittsburgh, PA 15213, USA
ycai@cmu.edu, sihan416@gmail.com

[2] SEI, Carnegie Mellon University, 4500 Fifth Ave, Pittsburgh, PA 15213, USA
jose@josemorales.org, austincasey@gmail.com

[3] Northrop Grumman Corporation, 1550 W Nursery Rd, Linthicum Heights, MD 21090, USA
neta.ezer@ngc.com

Abstract. Cyber malware has evolved from simple hacking programs to highly sophisticated software engineering products. Human experts are in high demand but are busy, expensive, and have difficulty searching through massive amount of data to detect malware. In this paper, we develop algorithms for machines to learn visual pattern recognition processes from human experts and then to map, measure, attribute, and disrupt malware distribution networks. Our approach is to combine visualization and machine vision for an intuitive discovery system that includes visual ontology of textures, topological structures, traces, and dynamics. The machine vision and learning algorithms are designed to analyze texture patterns and search for similar topological dynamics. Compared to recent human-machine teaming systems that use input from human experts for supervised machine-learning, our approach uses fewer samples, i.e. less training, and aims for novel discoveries through human-machine teaming.

Keywords: Visualization · Malware · Malware distribution network · Human-machine teaming · Machine learning · Computer vision · Pheromone · Security · Dynamics · Graph

1 Introduction

Over the past two decades, malicious software have evolved from simple hacking programs to highly sophisticated software engineering products. Prevailing data-driven machine-learning methods, such as signature recognition and behavior profiling are not sufficient to discover and attribute the malware with advanced stealthy functions such as polymorphism and distributed hosting. In addition, rapidly growing malware contains many unknown species that, given a lack of training samples, makes traditional machine learning algorithms insufficient. Human experts are still in high demand in this area because they often use intuition, instincts, and "non-linear" thinking strategies to discover unknown classes of malware and unknown distribution patterns, which

T. Ahram and W. Karwowski (Eds.): AHFE 2019, AISC 960, pp. 97–108, 2020.
https://doi.org/10.1007/978-3-030-20488-4_10

existing autonomous malware discovery systems are not capable of matching. However, experts are busy and expensive. We need "one-shot" or even "zero-shot" machine-learning algorithms to detect, map, and predict malware types and distribution networks. Here we aim to develop an *artificial intelligence virtual analyst* that enables a machine to learn cognitive discovery processes from human experts and then to map, measure, attribute, and disrupt malware distribution networks.

A malware distribution network (MDN) is a set of top level domains (TLD) that have been maliciously compromised. The TLD nodes are connected with directed edges where the direction indicates the flow of malicious traffic across domains. These connected nodes create a hidden structure used by malicious authors to spread malware globally. MDNs are often used as a backend distribution network for attack campaigns such as botnets, spams, and distributed denial of service attacks [1–4]. MDN is one of the key financial sources of underground activities, malicious actors, and the overall hidden economy [3]. MDNs are highly dynamic with domains and associated IP addresses changing constantly. Detection of MDN is normally based on intensive manual analysis of log files, identifying the malicious nodes, placing the appropriate edges, and plotting graphs. Usually, an MDN won't be discovered until the overall network distribution graph is plotted.

The data for this work is based on the Google Safe Browsing transparency report (GSB) [5] along with malicious attribution provided by VirusTotal online service. This work combines GSB and VirusTotal data to construct malware attributed MDNs over the 9-month data collection period. The main contributions of this work include: (1) a novel human-machine teaming model revealing existence of highly persistent sub-MDN networks from the cloud-sourcing dataset, and (2) a combination of network visualization, machine vision, and machine learning for dynamic data analytics.

2 Related Work

MDNs can be found in URLs [6, 7], botnets, pay-per-install, traffic redirection and manipulation [8–10]. Provos et al. [11, 12] show that the existence of MDNs web-based malicious campaigns, which led the launch of GSB service. In this study, we enhance our previous work in visual analytics [21, 24] using a "breadcrumb-like" pheromone model which traces topological changes over time.

In the wake of growing demand for human-machine teaming technologies for cyber security, in April, 2018, DARPA announced the program CHESS (Computers and Humans Exploring Software Security), which aims to develop human-machine teaming system for rapidly discover all classes of vulnerability in complex software. In contrast to the previous DARPA Cyber Grand Challenge (CGC) program that focuses on autonomous systems, CHESS program includes humans in the discovery processes in order to overcome the weaknesses of fully automated discovery systems.

MouselabWEB [8] is a process-tracing tool for extracting the process of decision makers. The technology was originally developed for understanding the decisions made by financial analysis, but has been used in other applications in use by consumers. MouselabWEB has its disadvantages. For example, it is limited by predetermined structures and orders of input as it was designed for business decision-making in general. Thus, it does not have the flexibility for enabling the discovery process used by cyber analytic experts.

The Renaissance of ontologies is attributed to the rise of *Explainable AI* as a way to provide insight rather than just information. Many graphical user interfaces visualize ontologies as a graph that reflects relationships among concepts. Visual ontology, on the other hand, is broader than just visualization. It contains more explicit cognitive content to enable users to intuitively understand the relationship between concepts and the overall knowledge structure with *psycho-physical* meanings such as distances, attractive forces, and repulsive forces, pheromones, and dynamics.

There are a few UML-based ontology visualizations [9–11, 16]. These existing visualization systems for ontologies have several weaknesses: (1) they are static, failing to represent dynamic information or changes over time; (2) the representations are limited within 2D diagrammatic views without multimodal inputs such as video, audio, and other sensory signals; (3) most entries are manual without connecting to autonomous machine learning systems, nor real-time data collection systems; and (4) they are limited at their language level, rather than broader domain applications, such as Internet of Things (IoT), cyber vulnerability, or healthcare.

T-SNE (t-Distributed Stochastic Neighbor Embedding) is a method for dimensionality reduction used for analyzing extremely high-dimensional datasets [12]. It is a promising solution for conventional methods such as Force-Directed Graph are computationally expensive (NP-Hard Problems) and cannot process large graph datasets.

There are also many visual simulation models such as continuous system dynamics language DYNAMO and discrete system dynamics simulation model Petri Nets. Both can be computerized. However, the models are not scalable to broader applications.

Finally, computer vision has been used to classify malware types [13]. By transforming the binary data into a grey-level image, texture features can be extracted and then classified with learning algorithms. This approach can help human analysts to visually understand and validate the patterns of malware. It also helps human analysts create novel visual ontologies of malware families and their behaviors. This is useful during initial data exploration without any prior knowledge about the dataset. However, features other than texture descriptors may get lost in the transformation. In fact, this is a dimension reduction method that may not work for every dataset.

3 Human-Machine Visual Learning Model

Our human-machine teaming discovery system contains three components: (1) cyber crawlers for data collection, (2) visual ontology for visualization and pattern extraction, and (3) human-machine collaborative learning. Figure 1 illustrates the architecture of the system.

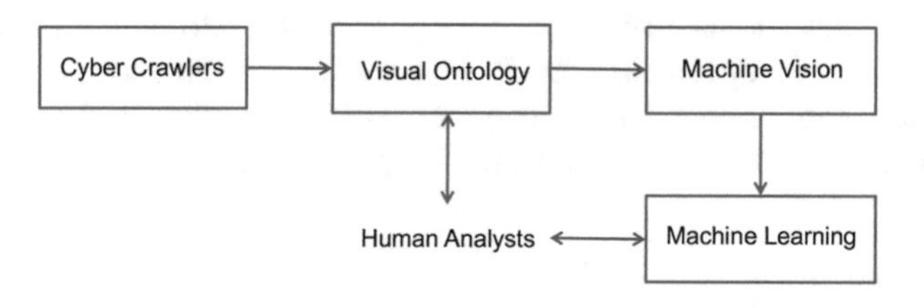

Fig. 1. Architecture of the human-machine teaming discovery system

The cyber crawlers are designed to continuously collect data on malware patterns, distribution, and attributions. The data is stored in JSON format. The dataset is converted to texture images and graphs that reflect the changes over time. The visual patterns are represented in visual ontology for human analysts to observe and for machine vision to perceive automatically, including anomaly detection and clustering. Finally, machine learning algorithms help human analysts to find similar dynamic patterns in sub-graphs and computer the distances based on similarities.

4 Cyber Crawlers for Data Collection

In this study, we scripted crawlers to map malware attribution and malware distribution network (MDN) data from available sources such as Google Safe Browsing (GSB) and VirusTotal. The GSB reports contain timestamps, network nodes and links, while VirusTotal reports contain malware type and submission timestamps. The crawler first performs the data integrity checking by mapping abstract data into multi-layer textures and detect anomalous events based on texture classification models such as energy, contrast, and entropy. Then the crawler builds a malware distribution graph to show which nodes are infected by which malware, their sources and destinations with timestamps. This can help the analysts grasp a big picture of the malware distribution and their evolutionary dynamics, including the sizes of the malware distribution clusters, root nodes of malware host, super-nodes in the malware distribution network, persistent links during the distribution, and the minimal-cut strategies to disrupt the malware distribution network. Figure 2 illustrates the MDN generation process.

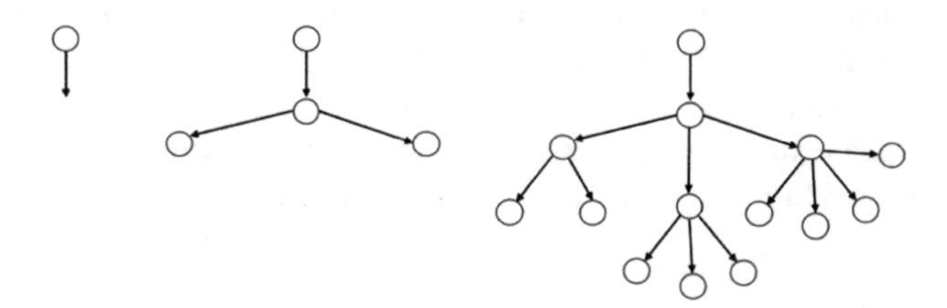

Fig. 2. Generating malware distribution network (MDN) from crawlers over time

Our 9 month collection period occurred from 19 January to 25 September 2017. Querying Google Safe Browsing (GSB) Transparency Report served as our main source of data. The process was seeded by requesting a GSB report for the often detected as malicious domain *vk.net* once every 24 h. Our selection of *vk.net* was based on the sites consistent appearance on GSB over a four-month period. This site would return a report that provided links to other TLDs that were victims of infection, meaning those sites received malicious traffic from our seed site.

5 Visual Ontology for Pattern Representation

Visual ontology in this study includes two layers of descriptions: First, at topological level, e.g. the Top Level Domain Names and their graph connections; second, at the binary code level, we will map the bits into texture and color patterns so that *both* human analysts and computers can see it. Figure 3 shows visual ontology (in texture and color) of the data from a mobile phone. After converting abstract data into visual artifacts, we then use machine vision and learning algorithms to recognize data structures, malware types, and topological dynamics Figs. 4 and 5.

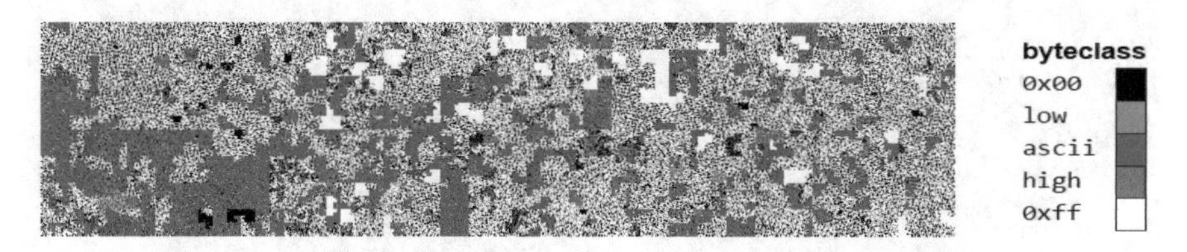

Fig. 3. Visualization of an overview of extracted data from a mobile phone

In this study, we defined a set of visual objects as a language for describing patterns and trends in the data. These visual objects include: (1) *texture* – mapping data to an image with variation of intensity or color, (2) *topology* – graph relationship of a connected network, (3) *trace* – duration of interaction between two nodes (entities), and (4) *dynamics* – changes of state over time (e.g. texture, topology and trace).

We tested the texture pattern mapping with our collected data from Google Safe Browsing (GSB). Each website entry in a captured collection has 24 base attributes which GSB provides. We additionally calculate 2 attributes, root malicious and web site type. Root malicious is a binary attribute specifying whether there are any inbound edges for the website. Website types are determined by the label GSB assigns, which is an attack site, intermediate site, or malicious node (default label when no other label is assigned). Each non-quantitative value is valued either by assigning numerical category value or by the attribute length. Then each value is normalized to be between 0 and 1 by dividing each attribute value by the maximum value across any collection. Each site entry is then visualized in a 6×5 patch of 5×5 pixel squares. Each value is converted to a brightness value for a white color and rendered into an image.

In the general data, the data quantity in the collection rarely fall below one quarter of the largest collection we had across the months. The two other groups of raw data are low quantity data. Some data collections are significantly less than others, but nevertheless cover a substantial group of websites. We believe these are anomalous data sets given there is a sharp discontinuity in attributes. These corresponding collections may have unique traits which are worth further investigation. Lastly we have the known incomplete data sets, which may have occurred due to script crashes, computer outages or any other external reason which generated near empty data sets. This visualization helps us weed out those sets in an effective visual manner.

Fig. 4. Examples of visualization of five data entry (zoom-in)

Fig. 5. Types of textural patterns of collections: normal data (left), anomalous data (middle) and incomplete data – no data or very few entries (right)

Topology is an important visual language that can be used by humans and computers. A graph is an abstraction of relationships between entities. Figure 6 shows typical topological patterns of a graph. In this project, we continued our development of a 3D graph visualization algorithm with pheromone-based trace model and dynamic visualization for the GSB dataset of malware distribution networks.

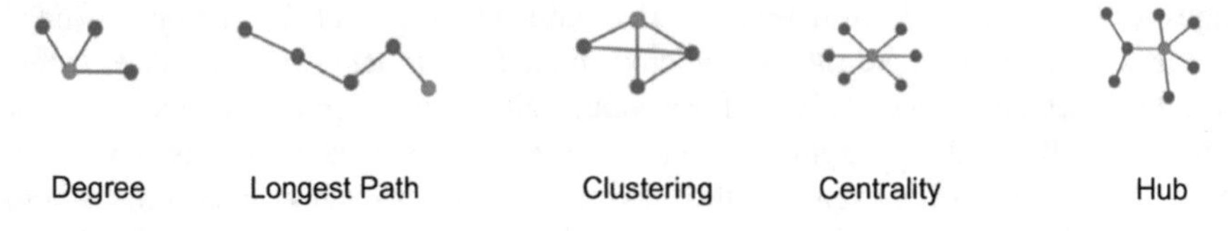

Fig. 6. Topological patterns of graph

In this study, we describe the MDN data structure as a list timestamp, source-destination network node pairs with associate attributes. The computer reads the dataset row by row to form the dynamic graph, where the distances between nodes are articulated with semantic attributes, for example, if the two nodes interact each other frequently, they would move closer and closer until they reach a collision limit. We also use the pheromone model [14] to visualize the persistency of malware on MDN edges and nodes, where pheromone deposit represents how long the malware stays. The longer the malware stays, the darker or bigger the shape. Pheromone decay represents that the impact on the edge or node becomes lighter and smaller [15]. See Fig. 7.

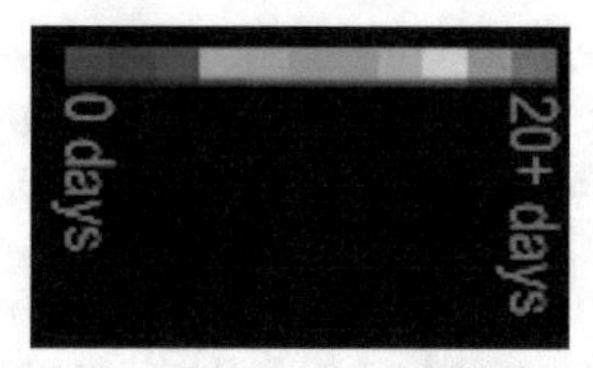

Fig. 7. The color map of the digital pheromone representation of persistency of the malware distribution over the connected links (edges), ranging from 0 day to 20 days or more

The visualization of the malware helps the cyber analysts to discover the dynamics of MDN clustering, for example, several large clusters grew in one to two weeks. Smaller chains first formed over time before multiple chains gathered together to form a large cluster. From our dataset in 2017, we observed a large centralized cluster of similar domains in February, March, and June. This indicates potential large scale malware attacks during those periods.

From visualization, we also observed the correlation between social events and formation of large clusters of MDNs. For example, ten days after the US Presidential Inauguration Day, January 20, 2017, there was a significantly large cluster of MDN that continued till middle of February. See Fig. 8.

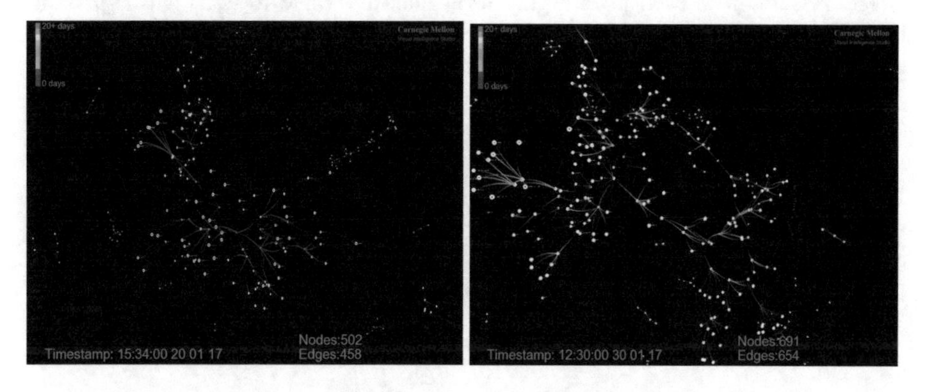

Fig. 8. A cluster of MDN on the US Presidential Inauguration Day, January 20 (left) and the cluster formation on January 30 (right), 2017

6 Human-Machine Collaborative Learning Model

In this study, we developed a novel machine learning algorithm that uses visual ontology (e.g. texture and topology) as "feature vectors", or hidden layers in a neural network. It has more semantic knowledge, or augmented human intuition, than purely numerical learning. Given samples of data, we can generate sequences of events in the real-world environment. The input and output data can be used to train a machine learning algorithm to recognize significant patterns. To build the model, we need to extract features. For example, we can map the raw malware data to textured images and apply computer vision models to characterize statistical texture attributes in small patches, such as variance, uniformity, flatness, and entropy. We then train the machine-learning model with known data structures and texture patterns. We anticipate that the graph and texture feature representations can bridge the cognitive perception, and computerized pattern recognition to improve the chance of data recovery, improve the processing speeds, and move toward an automated analysis process.

6.1 Texture Perception Model

A goal of the texture perception model was to train a machine to see texture patterns like humans while discovering attributes that are difficult for the human visual perceptual system to detect. We used the texture descriptors to cluster the data with 24 attributes. This method is different from the prevailing machine learning algorithms that map data to an abstract space without visual explanation. To easily categorize the data sets, we plotted each three choice combinations from the four texture descriptors (entropy, variance, uniformity, and flatness). Each image in grayscale has each pixel location grayscale value normalized into a probability. The values used for the following calculations are probability scaled. Originally, each pixel contains a grey scale value that is between 0 and 1. The probability-scaled values are new values for each pixel that is the pixel grayscale value divided by the summed grayscale values for the image. Assume p is each pixel's gray level intensity value (0–255) and $H(p)$ is the histogram. $N(p)$ is an estimate for the probability of any given pixel intensity value occurring in the image: $N(p) = \mathrm{H}(p)/|H(p)|$ and m is the mean of the probability of any given pixel intensity value occurring in the image. We have:

$$\text{variance} = \sum_{i=1}^{255} (p_i - m)^2 \cdot N(p_i)$$

$$\mathit{flatness} = \sum_{i=1}^{255} (p_i - m)^4 \cdot N(p_i)$$

$$\mathit{uniformity} = \sum_{i=1}^{255} N^2(p_i)$$

$$\mathit{entropy} = -\sum_{i=1}^{255} N(p_i) \cdot \log_2 N(p_i)$$

We used the first order texture descriptors instead of the second order descriptors of the grey-level co-occurrence matrix (GLCM) because the GLCM model is sensitive to the arrangement of the neighborhood of the data. In our case, the order of the mapping sequence is less important. From Fig. 9, we can easily distinguish anomalous data from normal data.

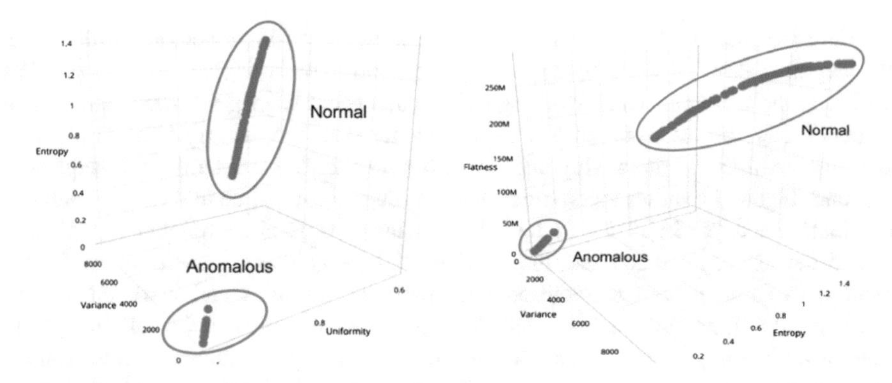

Fig. 9. Texture analysis of GSB dataset in entropy, variance, flatness, and uniformity

6.2 Search for Similar Topological Dynamics

How can we train a machine to find similar dynamic patterns? In this study, we focus on tracking topological pattern changes over time and finding similar patterns in the dataset. Figure 10 shows an example of three simplified malware distribution networks that changed over a period of three days. The topological changes can be represented by a set of graphic measurements such as graph degree. From the figure we see that the first and the third graphs are similar in changing patterns, although the slopes of the changes are slightly different.

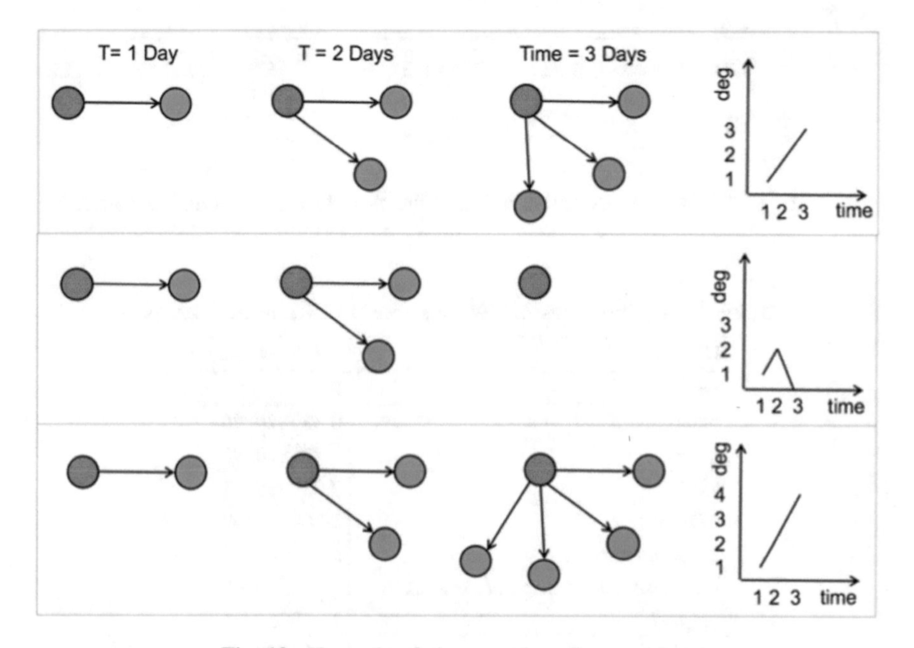

Fig. 10. Example of changes of graphs over time

Here we consider two sequences of the topological attribute vectors, in this case, the degree values: $A = a_1, a_2, a_3, \ldots, a_i, \ldots a_n$ and $B = b_1, b_2, b_3, \ldots, b_j, \ldots b_m$. We start with the Dynamic Time Warping (DTW) algorithm to map the two sequences of feature vectors on two sides of a two-dimensional grid. We then try to align them by warping the time axes iteratively until the shortest distance between the two sequences is found. Figure 11 shows topological changes (degree) over time in a sample malware distribution network in 2017 and the DTW distances between the reference graph *goo.gl* and the rest. As we can see, the *bambaee.tk* is the closest one. In this study, we explore dynamic pattern recognition with multiple attributes and work on general matching algorithms that can be applied to specific case studies in the following task. The visual learning algorithm can be used for searching for similar dynamic patterns and can also be used for self-organized clustering for exploring new visual ontological classes or trends. See Fig. 11 and Table 1.

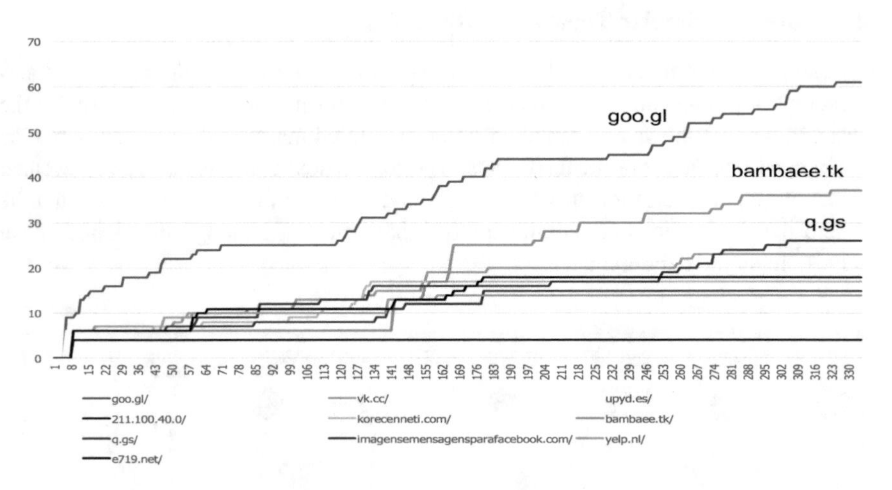

Fig. 11. DTW distance calculation indicates the closest two: *goo.gl* and *bambaee.tk*

Table 1. Inquiry results in DWT distances to URL address *goo.gl*

URL/IP address	DTW distance
goo.gl/	0
bambaee.tk/	0.335820896
q.gs/	6.735820896
vk.cc/	7.832835821
e719..et/	10.00746269
korecenneti.com/	10.41044776
imagensemensagensparafacebook.com/	11.48358209
yelp.nl/	11.69253731
upyd.es/	14.58358209
211.100.40.0/	16.40447761

7 Conclusions

In this study, we explore a human-machine teaming discovery system that combines visualization, machine vision, and learning algorithms for discovering dynamic topological patterns from a real-world dataset. We use cyber crawlers to fetch Malware Distribution Network (MDN) data and convert the dataset into a texture and graph. The visualization of MDN with pheromone deposition and decay reveals its topological structure, persistence, and evolution over time.

We found that the texture mapping enables both humans and machine vision algorithm to cluster patterns in the raw data and spot anomalies and missing data. From the human-machine discovery process, we are able to search subgraphs that have similar topological dynamics. Furthermore, we are able to discover the critical MDN infrastructural elements such as hubs and bridges. With search algorithms for those elements, the computer is able to mimic human analysts' visual intuition to locate pivotal hubs and bridges for timely response. The approach enables a human-machine teaming approach for novel malware discovery and understanding.

Acknowledgement. The authors would like to thank research assistants Pedro Pimentel and Sebastian Peryt for early prototypes. This project is in part funded by Cyber-Security University Consortium of Northrop Grumman Corporation. The authors are grateful to the support from Drs. Paul Conoval, Robert Pipe, and Donald Steiner. [DISTRIBUTION STATEMENT A] This material has been approved for public release and unlimited distribution. Please see Copyright notice for non-US Government use and distribution. Internal use: * Permission to reproduce this material and to prepare derivative works from this material for internal use is granted, provided the copyright and "No Warranty" statements are included with all reproductions and derivative works. External use: * This material may be reproduced in its entirety, without modification, and freely distributed in written or electronic form without requesting formal permission. Permission is required for any other external and/or commercial use. Requests for permission should be directed to the Software Engineering Institute at persmission@sei.cmu.edu. * These restrictions do not apply to U.S. government entities. Carnegie Mellon® and CERT® are registered in the U. S. Patent and Trademark Office by Carnegie Mellon University. DM19-0291.

Distribution Statement A: Approved for Public Release; Distribution is Unlimited; #19-0490; Dated 04/17/19.

References

1. Gu, G., Perdisci, R., Zhang, J., Lee, W.: BotMiner: clustering analysis of network traffic for protocol- and structure-independent botnet detection. In: Proceedings of the 17th USENIX Security Symposium (Security 2008) (2008)
2. Gu, G., Zhang, J., Lee, W.: BotSniffer: detecting botnet command and control channels in network traffic. In: Proceedings of the 15th Annual Network and Distributed System Security Symposium (NDSS 2008) (2008)
3. McCoy, D., et al.: Pharmaleaks: understanding the business of online pharmaceutical affiliate programs. In: Proceedings of the 21st USENIX Conference on Security Symposium, Ser. Security 2012, pp. 1–1. USENIX Association, Berkeley (2012)

4. Karami, M., Damon, M.: Understanding the emerging threat of ddos-as-a-service. In: Proceedings of the USENIX Workshop on Large-Scale Exploits and Emergent Threats (2013)
5. Google safe browsing. https://developers.google.com/safe-browsing/
6. Peryt, S, Morales, J.A., Casey, W., Volkmann, A., Cai, Y.: Visualizing malware distribution network. In: IEEE Conference on Visualization for Security, Baltimore, October 2016
7. https://ieeexplore.ieee.org/stamp/stamp.jsp?arnumber=7870760
8. MouselabWEB. www.mouselabweb.org
9. Barzdinš, J., Barzdinaš, G., Cerans, K., Liepinš, R., Sprogis, A.: OWLGrEd: a UML style graphical notation and editor for OWL 2. In: Proceedings of the 7th International Workshop on OWL: Experiences and Directions (OWLED 2010), volume 614 of CEUR-WS (2010)
10. Kost, R.: VOM – Visual Ontology Modeler (2013). thematix.com/tools/vom/
11. Howse, J.: Visualizing ontologies: a case study. In: International Semantic Web Conference, ISWC 2011, pp. 257–272. Springer, Berlin (2011)
12. van der Maaten, L.J.P., Hinton, G.E.: Visualizing non-metric similarities in multiple maps. Mach. Learn. **87**(1), 33–55 (2012)
13. Nataraj, L., et al.: Malware Images: Visualization and Automatic Classification, VizSec 2011, 20 July 2011
14. Cai, Y.: Pheromone-based visualization model of malware distribution networks. In: International Conference on Computational Science, to appear on Springer LNCS (2018)
15. Wigglesworth, V.B.: Insect Hormones, pp. 134–141. WH Freeman and Company, Holtzbrinck (1970)
16. TopQuadrant. TopBraid Composer. http://www.topquadrant.com/topbraid/

HackIT: A Human-in-the-Loop Simulation Tool for Realistic Cyber Deception Experiments

Palvi Aggarwal[1(✉)], Aksh Gautam[2], Vaibhav Agarwal[2], Cleotilde Gonzalez[1], and Varun Dutt[2]

[1] Dynamic Decision Making Lab, Carnegie Mellon University, Pittsburgh, USA
palvia@andrew.cmu.edu, coty@cmu.edu
[2] Applied Cognitive Science Laboratory, Indian Institute of Technology Mandi, Mandi, India
{aksh_g, vaibhav_agarwal}@students.iitmandi.ac.in, varun@iitmandi.ac.in

Abstract. Deception, an art of making someone believe in something that is not true, may provide a promising real-time solution against cyber-attacks. In this paper, we propose a human-in-the-loop real-world simulation tool called HackIT, which could be configured to create different cyber-security scenarios involving deception. We discuss how researchers can use HackIT to create networks of different sizes; use deception and configure different webservers as honeypots; and, create any number of fictitious ports, services, fake operating systems, and fake files on honeypots. Next, we report a case-study involving HackIT where adversaries were tasked with stealing information from a simulated network over multiple rounds. In one condition in HackIT, deception occurred early; and, in the other condition, it occurred late. Results revealed that participants used different attack strategies across the two conditions. We discuss the potential of using HackIT in helping cyber-security teams understand adversarial cognition in the laboratory.

Keywords: Cybersecurity · Simulation tools · Learning · Attack · Hackers · Defenders · Honeypots

1 Introduction

Deception is an art of making someone believe in something that is not true, may provide a promising real-time solution against cyber-attacks [1]. Deception involves interaction between two parties, a target and a deceiver, in which the deceiver effectively causes the target to believe in a false description of reality [1]. The objective is to cause the target to work in such a way that is beneficial to the deceiver. Deception has been used as an offensive and defensive tool in cyber world by hackers and defenders. Hackers used deception for exploiting cyber infrastructure, stealing information, making money and defaming people. The deception techniques used by hackers may involve malware signature, conceal code and logic, encrypted exploits, spoofing, phishing, and social engineering (e.g., by deceiving help desk employees to install

T. Ahram and W. Karwowski (Eds.): AHFE 2019, AISC 960, pp. 109–121, 2020.
https://doi.org/10.1007/978-3-030-20488-4_11

malicious code or obtain credentials). However, defenders used deception for securing network infrastructure, luring hackers to fake information and understanding hacker's motives and strategies using tools such as honeypots. When used for defense, cyber defenders may use feints and deceits to thwart hackers' cognitive processes, delay attack activities, and disrupt the breach process. When used for defense, deception may be achieved through miss-directions, fake responses, and obfuscations [2]. These techniques rely upon hacker's trust in response from network, data, and applications during an actual attack. To create deception for defense, security experts have been using honeypots, fake servers that pretend to be real, for gathering intelligence about hackers. Honeypots are one of the effective deception tools in the network defense to lure hackers.

Decisions-making process of hackers and defenders in cyber world is complex task. Understanding their decision process in such complex environment is challenging. Simulation has been used as an effective way of understanding the hackers' and defenders' decisions, testing new solutions for security, and training the models and people in such complex task scenarios [3–10] have used behavioral game theoretic approaches to understand the decisions of hackers and defenders in the abstract cyber-security games. In such game theoretic approaches, the network structure and the set of actions of hackers and defenders were abstracted to attack/not-attack and defend/not-defend. Furthermore, the information provided to the participants was also abstracted. The task simulated using game-theoretic approaches was less cognitively challenging compared to the real cyber-security tasks. Thus, the conclusions made based on behavioral game-theoretic approaches may or may not address the cybersecurity problems. Another approach used to study hackers and defenders behaviour involve complex real time tools such as NeSSi, Network Simulator-2/3, Cymulate etc. [11, 12]. However, using deception in uncontrolled environments makes it difficult to answer cause-effect questions.

Aggarwal et al. [6] proposed HackIT tool to bridge the gap between behavioral game-theoretic approaches and real-world cybersecurity tasks. HackIT tool provided features to create more specific set of actions and information needed for cyber-security tasks for both hackers and defenders. The HackIT tool was used to replicate the results of a laboratory experiment using a deception game [3, 4]. Results revealed that the average proportion of attacks was lower and not-attack actions were higher when deception occurred late in the game rather than earlier; and when the amount or deception was high compared to low [6]. This result found in an abstract simplified scenario was replicated in a real-world simulation tool called the HackIT.

In [6], the HackIT tool was available for creating deception with limited number of systems only and for single player games only. In this paper, we define the enhanced capabilities of HackIT tool. Specifically, we detail how the HackIT tool is capable of running experiments with different sized networks, different network configurations, different deception strategies, and single player and multiplayer games. Thus, the enhanced capabilities of HackIT tool can help us answer several questions such as the effect of different network sizes and honeypot allocations on hacker's performance and the most effective way to present the "clues" of deception in network nodes.

In what follows, we first discuss the functioning and different features in the of the HackIT tool. Next, we detail an experiment involving the HackIT tool to showcase its capabilities. Furthermore, we detail the results from the experiment and discuss the implications of our results for investigating the decision-making of hackers in the real world.

2 HackIT

HackIT is a generic framework for cybersecurity tasks to study human learning and decision-making of hackers and analysts. It represents a simplified framework containing the most essential elements for creating cybersecurity scenarios: network nodes, which represents the characteristics of real nodes; strategies, which can be configured for creating deception; and commands, which are used for communication with the network. The analyst's goal in the HackIT task is to evaluate different deception strategies and the hacker's goal to identify the real network nodes and exploit them. Hackers communicate with the network in HackIT using different commands and gain information about different configurations. However, hackers are not aware of the strategies used by analysts inside the HackIT scenario. Hackers basically learn about these strategies overtime by playing different rounds. Thus, HackIT is a controllable and flexible simulation tool with the capability of creating various network scenarios and experiment with different techniques to lure hackers. Figure 1 shows the typical flow of HackIT tool which uses the concept of stackelberg security games [13], where first defenders create a network and use their defense techniques and next, hackers try to exploit the security of the network.

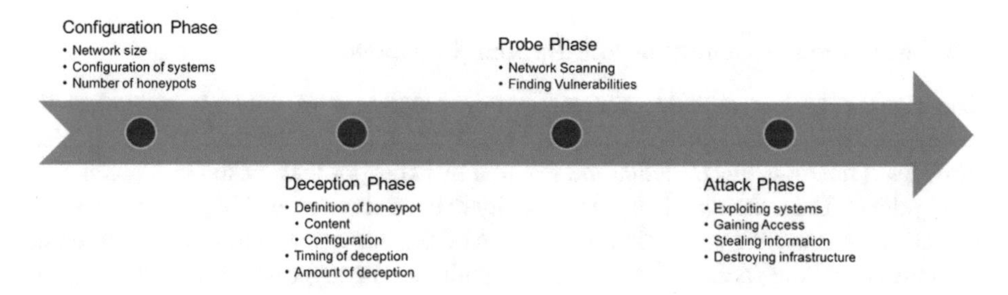

Fig. 1. HackIT tool

HackIT is an experimental cybersecurity tool that allow analysts to simulate cyber infrastructure during configuration phase and define the deception algorithm in deception phase. Next, these scenarios are presented to the hackers for exploitation. Hackers use this tool in two phases: probe phase and attack phase. The probe phase involves the process of reconnaissance where hackers gather information about open port, services, operating systems, and available vulnerabilities using tools such as nmap; whereas, the attack phase involves gaining access to different computers and stealing information or compromising computer systems.

Despite the simplicity of HackIT, the tool has the potential to simulate many real-world dynamic situations in the laboratory: testing different proportion of honeypots in the network, testing optimal placement of honeypots, different configurations of honeypots, such as, easy to exploit ports on honeypots, and availability of fake files on honeypots, availability of unpatched ports on honeypot etc. Specific characteristics of HackIT tool are explained as followed:

2.1 Network Size

The HackIT tool is flexible to create any number of computers in the network. The configuration of these systems will also be dynamically created. Thus, this tool can allow researchers to easily run experiments with small, medium and large scaled networks as shown in Fig. 2.

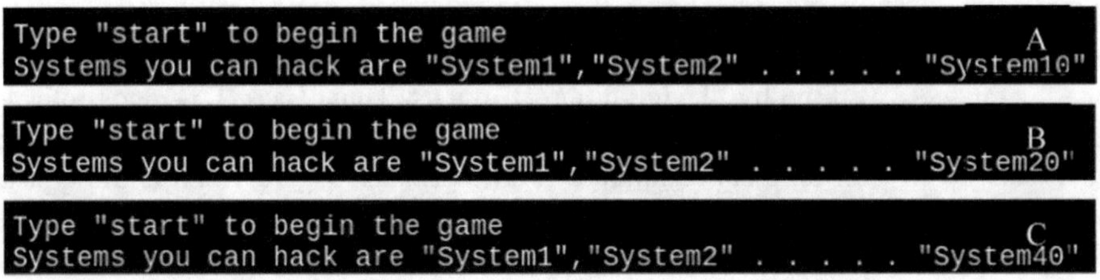

```
Type "start" to begin the game                                           A
Systems you can hack are "System1","System2" . . . . . "System10"
```

```
Type "start" to begin the game                                           B
Systems you can hack are "System1","System2" . . . . . "System20"
```

```
Type "start" to begin the game                                           C
Systems you can hack are "System1","System2" . . . . . "System40"
```

Fig. 2. Different network sizes: (A) Small, (B) Medium, and (C) Large

2.2 Allocation of Honeypots and Regular Computers

The HackIT tool is capable of creating any proportion of network computers as honeypots, where honeypots are fake computer pretending to be real. This tools also provides a functionality to define the features of honeypot that make them pretend as real systems. These features include the ability to configure vulnerable ports, vulnerable operating systems, and proportion of fake files on the honeypots. Thus, setting different proportion of honeypots and defining deception via honeypots is relatively easy in HackIT.

2.3 Configuration of Computers

The configuration of honeypots and regular computers is automatically generated by a script in HackIT that would produce the configuration for a fixed number of systems with a given proportion of honeypots and real systems. This script generates the configuration consisting of real systems, honeypots, and files for each game round in HackIT. By using this script, one could generate data onetime and encode it in the experiment so that it would present all participants with same configuration. For example, the regular systems could be configured as patched and difficult to exploit and honeypot systems could be configured as easy to exploit. This configuration mapping is shown in Table 1.

Table 1. Configuration of Honeypot and regular systems

Easy to attack	Difficult to attack
Operating systems: • Windows Server 2003 • Windows XP • HP-UX 11i • Solaris	Operating systems: • OpenBSD • Linux • Mac OS X • Windows 8
Services and ports: • 21/tcp – ftp • 25/tcp – smtp • 80/tcp – http • 111/tcp – rpcbind • 135/tcp – msrpc	Services and ports: • 22/tcp-ssh • 53/tcp-domain • 110/tcp-pop3 • 139/tcp-netbios • 443/tcp-https • 445/tcp-microsoft-ds • 3306/tcp-mysql • 5900/tcp-vncc http • 6112/tcp-dtspc • 8080/tcp-apache

2.4 Content of Honeypots

HackIT tool provide a facility to create fictitious content on honeypots using fake files. The proportion of fake files and useful files on a honeypot can be configured in HackIT. Figure 3 shows the output of the *ls* command in the directory where only pin.txt is a real file and rest are fake files. In HackIT tool, the number of files on each server can be dynamically created. We tested our platform with the different number of files ranging from 50–200.

```
You have gained entry into the system. Now use ls to see the files in the system

> ls

pin.txt
cras_in_purus.txt
amet.txt
ultrices_erat_tortor.txt
at_velit_vivamus.txt
eleifend_luctus.txt
nullam.txt
nullam_varius.txt
tellus.txt
ut_nulla_sed.txt
orci_mauris.txt
quam_nec.txt
est.txt
```

Fig. 3. List of fake files on Honeypot webserver

2.5 Single-Player and Multi-Player Platform

HackIT tool provides a functionality to run single-player and multi-player experiments. In single-player experiment settings, players cannot communicate with each other.

However, in multi-player experiment setup, players are provided with a chat functionality to share their knowledge as shown in Fig. 4. Hackers usually penetrate into the network by organizing themselves into a group. Hackers in a group rely on information gained from fellow hackers who have already penetrated into the network or are trying to penetrate it.

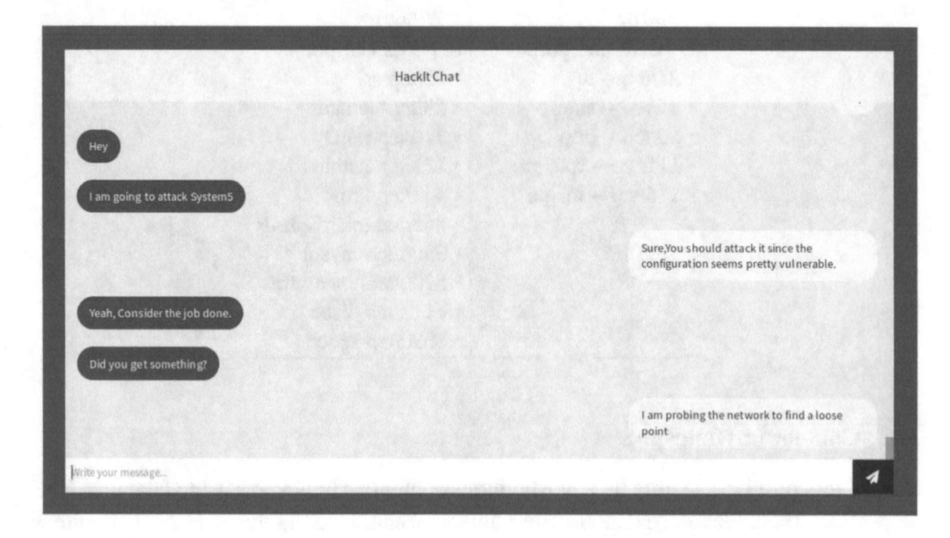

Fig. 4. Chat functionality for multiplayer experiments in HackIT

2.6 Commands

The HackIT tool can run various network commands that include: *nmap*, *use_exploit*, *ls*, and *scp*. *Nmap* is a network utility that shows the open ports, operating system type, and services running on the specified webserver. The *nmap* utility also provides the list of vulnerabilities on the corresponding webservers. The *use_exploit* command exploits vulnerabilities of a system and helps attacker to gain access to a particular webserver. Next, the *ls* command lists the files currently on the file system of the machine. The *scp* command transfers files to the remote machine.

2.7 Measures for Human-in-the-Loop Experiments

For understanding the decision-making process of hackers, it is important to analyze how do they gather information, their preferences while searching for information, kind of exploits used, and most importantly how to they react to different defense strategies. HackIT tool provides various measures to analyze the human performance in cyber-security experiments.

- **Probing behavior:** HackIT tool provides capability to record the probing behavior of hacker participants. Probing is an important phase where hackers collect information before launching a real attack. Using deception in probing responses and analyzing hackers' reaction towards deceptive responses is an important measure provided in HackIT tool.
- **Attack behavior:** HackIT measures the attack behavior of hackers by recording their attack actions, exploit information, their specific selection of targets or configuration. HackIT tool also records the vulnerabilities exploited, and exploits used by hackers. Defenders can analyze this data to study their attack patterns.
- **Learning:** HackIT tool can record overtime behavior of participants. Thus, it provides data to analyze learning capabilities of hackers against different deception strategies.
- **Communication History:** HackIT tool provides functionality to analyze the chatting history of hacker to investigate that how hackers make decisions or formulate strategies in team-based tasks.
- **Time:** HackIT tool also records time taken by hackers to exploit a system. Timing is another measure to evaluate the success of deception.

3 Case Study Using Hackit Tool

3.1 Experiment Design

In this paper, we demonstrated the ability of HackIT tool as a single-player platform for conducting experiments. We analyzed the effect of timing of deception on hacker's actions in a study. The timing of deception was manipulated across two between-subject conditions: early deception (N = 8) and late deception (N = 8). The total number of servers were 40, where 20 webservers were assigned as honeypot webservers during deception rounds. In both conditions, participants playing as hackers were given 6 game rounds in a sequence (end point unknown to participants), where 2-game rounds possessed deception. In this experiment, if the timing of deception was early, then deception was present on the second and third rounds in the sequence. However, if the timing of deception was late, then deception was present in the fourth and fifth rounds in the sequence (see Fig. 5). The honeypots were easy to exploit via common ports and vulnerabilities in the deception rounds compared to the non-deception rounds, where there were no honeypots (more details ahead in this paper). However, participants were not told that honeypots will involve easy to attack configurations in deception rounds. Also, participants were not disclosed the rounds on which deception was involved. To analyze human data, we looked at the proportion of honeypot attacks and proportion of regular attacks at the attack stage by the hacker across six-rounds in each condition. These proportions were calculated by average the attack decision over all the trials and participants. We also calculated frequency of each exploits used on regular and honeypots in deception and no-deception trials.

Condition/ Rounds	1	2	3	4	5	6
Early Deception	-	D	D	-	-	-
Late Deception	-	-	-	D	D	-

Fig. 5. Experiment design using deception game with six rounds and two conditions, i.e. early deception and late deception. D: deception present -: deception not present [1, 2].

3.2 HackIT Task

The objective of attacker in HackIT was to steal real credit-card information located on one of the webservers. As shown in Fig. 1, defender first configure the network with 40 webservers where 20 webservers acted as honeypots. Defender also set up a strategy where honeypots were easier to attack during the deception rounds. Based on these strategies, in this experiment, defender uses different configurations shown in Table 2. For example, a system with Windows XP operating system, port 80/tcp, and service http was easily exploitable. Such a configuration was mapped to a honeypot. However, a system with Linux operating system, port 22/tcp and service ssh was difficult to attack. Such a configuration was mapped to a regular webserver. Participants were informed about the easy to exploit and the difficult to exploit configurations in Table 2 as part of the instructions.

Table 2. Configuration of Honeypot and regular systems

Strategy	Operating system	Ports	Exploits
Honeypots: Easy to attack	Windows server 2003 Windows XP HP-UX 11i Solaris	21/tcp – ftp 25/tcp – smtp 80/tcp – http 111/tcp – rpcbind 135/tcp – msrpc	brute_force directory_harvest sql_injection DDoS_attack DoS_attack
Regular: Difficult to attack	OpenBSD Linux Mac OS X Windows 8	22/tcp-ssh 53/tcp-domain 110/tcp-pop3 139/tcp-netbios 443/tcp-https 445/tcp-microsoft-ds 3306/tcp-mysql 5900/tcp-vncc http 6112/tcp-dtspc 8080/tcp-apache	user_auth DNS_zone_transfer pop3_version DCOM_buffer_overrun drown_attack windows_null_session remove_auth remote_auth remote_exploit_buffer url_decoder

First, the attacker probed the network using nmap command to gain information about different webservers. Probing different webservers gave the information about the operating system, open ports, services, and vulnerabilities. The information provided to the attacker as a result of probing systems gave him an idea about the type of configuration on the probed system. Once the attacker collects information about open ports and services, he could attack a webserver by using the "use_exploit" command. The use_exploit command exploited vulnerabilities on a system and helped the attacker to gain access to that webserver. Next, the attacker could list different files on the exploited webserver by using the "ls" command. Next, the attacker could transfer required files containing credit card information (e.g., "pin.txt") using the "scp" command. After attackers copied the file from the exploited system, he was informed whether he was successful or not in stealing a real credit-card file from the computer via a text-based feedback.

3.3 Participants

Participation was voluntary and a total of 16 male participants participated in the study that was advertised via an email advertisement. Out of the 16 people, 62.5% people had taken a course in computer networks/security. The age of participants ranged from 18–22 years. About 56% were 3rd year and 44% were final year undergraduate students from Indian Institute of Technology Mandi, India. All the participants were remunerated at a fixed rate of INR50 for their participation in the study.

3.4 Procedure

Participants were given instructions about their objective in the HackIT task, and they were informed about their own action's payoffs. Specifically, human hackers were asked to maximize their payoff by stealing the real credit-card file from the network over several rounds of play (participants were not aware of the endpoint of the game). Each round had two stages: Probe stage and Attack stage. Hacker could probe the network using "nmap" utility in first stage. After probing the webservers, he received information about open ports, operating systems, services, and vulnerabilities associated with different webserver. Next, the hacker had to choose one webserver to exploit and exploit webservers using "use_exploit" command during attack stage. Once the webserver was exploited, hackers transferred the credit-card file to their remote computer.

3.5 Results

Figure 6 shows the proportion of attacks on honeypot and regular webservers. There was no difference in the proportion of attacks in late and early deception conditions.

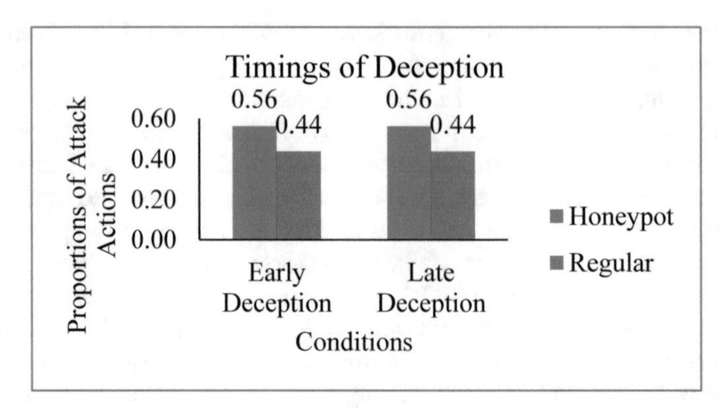

Fig. 6. Proportion of attack actions on regular and honeypot webservers across rounds and participants.

Next, we analyzed the exploits used during deception rounds and non-deception rounds by participants. When regular (honeypot) systems are attacked during deception rounds, that is called as deception failure (success). Figure 7a and b show the number of regular attacks and honeypot attacks against different exploits in deception failure and success, respectively. During deception failure, the remote_auth vulnerability was more exploited in early condition compared to late condition and the pop3_version vulnerability was exploited more in the late condition compared to early condition. During deception success, the brute_force vulnerability was more exploited more in early condition compared to late condition and the DOS_attack and sql_injection vulnerabilities were exploited more in the late condition compared to early condition.

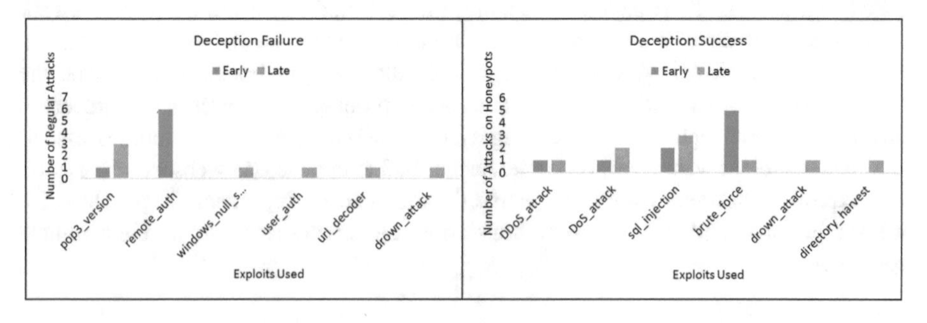

Fig. 7. Vulnerability exploited on regular systems and honeypot in deception rounds

Figure 8 shows the number of attacks on regular webservers using different vulnerabilities. We found that during early deception conditions, mostly hackers used remote_auth and drown_attack vulnerabilities. Furthermore, during late deception condition, hackers used pop3_version and remote_auth vulnerabilities.

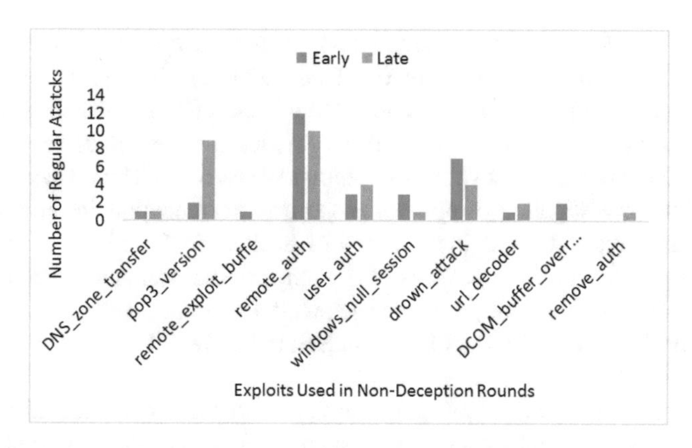

Fig. 8. Vulnerability exploited in non-deception rounds

4 Discussion and Conclusions

In this paper, we discussed HackIT, a HITL simulation tool with a potential to help cyber-security researchers to investigate the decision-making of attackers and defenders in real-world cyber-security scenarios. In this paper, we showed different features of HackIT tool and different ways to conduct multiplayer experiments. We also showed a concrete example of using HackIT to investigate the effects of timing of deception on hacker's decisions. We believe that HackIT tool would be helpful in creating other cyber-security scenarios involving dynamic network sizes, dynamic network configurations, and various deception strategies.

First, we simulated an information stealing scenario in HackIT. We found that the attacks on regular and honeypots were no different in early and late deception condition. One likely reason for this result could be that participants perceived easy to attack and difficult to attack vulnerabilities similarly. In fact, hacker participants exploited remote_auth vulnerability to attack on regular machines and brute_force and sql_injection vulnerabilities to attack honeypot systems. Furthermore, we found participants attacked more number honeypot systems compared to regular systems.

Cybersecurity faces different open challenges while implementing the deception. These challenges may involve the following questions: what an effective deception strategy should be? when should the deception be used? how hackers can be deceived during the probe phase? what are their probing patterns? and, how to make deception less detectable? HackIT tool could provide a framework to investigate these questions. One way to make deception less detectable is to have effective configuration and content on deceptive nodes. HackIT tool could be used to identify effective configurations and contents on honeypots to make them less detectable.

In future, we plan to perform a series of experiments involving participants performing as attackers in other simulated network scenarios in HackIT. Here, we wish to extend the HackIT tool to investigate the optimal proportion of honeypots and effectiveness of deception in networks of different sizes. For example, a network could be classified as small, medium, or large sized based on number of webservers present.

Furthermore, installing and maintaining honeypots is a costly affair for analysts in terms of money, time, and manpower. Thus, identify the optimal proportion of honeypots required in the network and testing their effectiveness in HackIT could reduce the overall cost. HackIT tool also provides the flexibility to configure any number of webservers as honeypots and regular webservers. Thus, HackIT tool could be useful in the investigation of optimal proportion of honeypots in the network. One could also investigate the probing patterns of hackers in HackIT. For example, one could investigate the probing patterns [14] such as local preference scanning, preference sequential scanning, non-preference sequential scanning, and parallel scanning in the human hacker's data in HackIT to understand different strategies for collecting information.

Furthermore, the HackIT tool could be useful to understand the effectiveness of deception against different cyber-attacks such as SQL injection, Denial of Service (DoS), and zero-day attacks. Overall, the HackIT tool provides a powerful platform to investigate various factors that could help analysts plan better cyber defense against hackers.

Acknowledgments. Research was partially sponsored by the Army Research Laboratory and was accomplished under Cooperative Agreement Number W911NF-13-2-0045 (ARL Cyber Security CRA). The views and conclusions contained in this document are those of the authors and should not be interpreted as representing the official policies, either expressed or implied, of the Army Research Laboratory or the U.S. Government. The U.S. Government is authorized to reproduce and distribute reprints for Government purposes notwithstanding any copyright notation here on. This research was also supported by the ICPS DST grant (T-533) from the Indian Government to Dr. Varun Dutt.

References

1. Whaley, B.: Toward a general theory of deception. J. Strateg. Stud. **5**(1), 178–192 (1982)
2. Rowe, N.C., Custy, E.J.: Deception in cyber attacks. In: Cyber Warfare and Cyber Terrorism, pp. 91–96. IGI Global (2007)
3. Aggarwal, P., Gonzalez, C., Dutt, V.: Cyber-security: role of deception in cyber-attack detection. In: Advances in Human Factors in Cybersecurity, pp. 85–96. Springer, Cham (2016)
4. Aggarwal, P., Gonzalez, C., Dutt, V.: Looking from the hacker's perspective: role of deceptive strategies in cyber security. In: 2016 International Conference On Cyber Situational Awareness, Data Analytics And Assessment (CyberSA), pp. 1–6. IEEE (2016)
5. Aggarwal, P., Gonzalez, C., Dutt, V.: Modeling the effects of amount and timing of deception in simulated network scenarios. In: 2017 International Conference On Cyber Situational Awareness, Data Analytics And Assessment (Cyber SA), pp. 1–7. IEEE (2017)
6. Aggarwal, P., Gonzalez, C., Dutt, V.: Hackit: a real-time simulation tool for studying real-world cyber-attacks in the laboratory. In: Handbook of Computer Networks and Cyber Security: Principles and Paradigms. CRC Press (in press)
7. Bye, R., Schmidt, S., Luther, K., Albayrak, S.: Application-level simulation for network security. In: Proceedings of the 1st international conference on Simulation Tools and Techniques for Communications, Networks and Systems & Workshops, p. 33. ICST (Institute for Computer Sciences, Social-Informatics and Telecommunications Engineering) (2008)

8. Maqbool, Z., Makhijani, N., Pammi, V.C., Dutt, V.: Effects of motivation: rewarding hackers for undetected attacks cause analysts to perform poorly. Hum. Factors **59**(3), 420–431 (2017)
9. Dutt, V., Ahn, Y.S., Gonzalez, C.: Cyber situation awareness: modeling detection of cyber-attacks with instance-based learning theory. Hum. Factors **55**(3), 605–618 (2013)
10. Aggarwal, P., Maqbool, Z., Grover, A., Pammi, V. C., Singh, S., Dutt, V.: Cyber security: a game-theoretic analysis of defender and attacker strategies in defacing-website games. In: 2015 International Conference on Cyber Situational Awareness, Data Analytics and Assessment (CyberSA), pp. 1–8. IEEE. (2015)
11. Cymulate- Breach and Attack Simulation. https://cymulate.com. Accessed 15 Feb 2019
12. Issariyakul, T., Hossain, E.: Introduction to network simulator 2 (NS2). In: Introduction to Network Simulator NS2, pp. 1–18. Springer, Boston (2009)
13. Tambe, M.: Security and Game Theory: Algorithms, Deployed Systems, Lessons Learned. Cambridge University Press, Cambridge (2011)
14. Achleitner, S., La Porta, T.F., McDaniel, P., Sugrim, S., Krishnamurthy, S.V., Chadha, R.: Deceiving network reconnaissance using SDN-based virtual topologies. IEEE Trans. Netw. Serv. Manag. **14**(4), 1098–1112 (2017)

Mathematical Model of Intrusion Detection Based on Sequential Execution of Commands Applying Pagerank

Cesar Guevara[1(✉)], Jairo Hidalgo[2], Marco Yandún[2], Hugo Arias[1], Lorena Zapata-Saavedra[3], Ivan Ramirez-Morales[3], Fernando Aguilar-Galvez[3], Lorena Chalco-Torres[3], and Dioselina Pimbosa Ortiz[3]

[1] Universidad Tecnológica Indoamérica, Ambato, Ecuador
{cesarguevara, cienciamerica}@uti.edu.ec
[2] Universidad Politécnica Estatal del Carchi, Tulcán, Ecuador
{jairo.hidalgo, marco.yandun}@upec.edu.ec
[3] Universidad Técnica de Machala, Machala, Ecuador
{mlzapata, iramirez, flaguilar, lchalco, dpimbosa}@utmachala.edu.ec

Abstract. Cybersecurity in networks and computer systems is a very important research area for companies and institutions around the world. Therefore, safeguarding information is a fundamental objective, because data is the most valuable asset of a person or company. Users interacting with multiple systems generate a unique behavioral pattern for each person (called digital fingerprint). This behavior is compiled with the interactions between the user and the applications, websites, communication equipment (PCs, mobile phones, tablets, etc.). In this paper the analysis of eight users with computers with a UNIX operating system, who have performed their tasks in a period of 2 years, is detailed. This data is the history of use in Shell sessions, which are sorted by date and token. With this information a mathematical model of intrusion detection based on time series behaviors is generated. To generate this model a data pre-processing is necessary, which it generates user sessions S_m^u, where u identifies the user and m the number of sessions the user u has made. Each session S_m^u contains a sequence of execution of commands C_n, that is $S_m^u = \{C_1, C_2, C_3, \ldots, C_n\}$, where n is the position in wich the C command was executed. Only 17 commands have been selected, which are the most used by each user u. In the creation of the mathematical model we apply the page Rank algorithm [1], the same that within a command execution session S_m^u, determines which command C_n calls another command C_{n+1}, and determines which command is the most executed. For this study we will perform a model with sb subsequences of two commands, $sb = \{C_n, C_{n+1}\}$, where the algorithm is applied and we obtain a probability of execution per command defined by $P(C_n)$. Finally, a profile is generated for each of the users as a signal in time series, where maximum and minimum normal behavior is obtained. If any behavior is outside those ranges, it is determined as intrusive behavior, with a detection probability value. Otherwise, it is determined that the behavior is normal and can continue executing commands in a normal way. The results obtained in this model have shown that the proposal is quite effective in the

T. Ahram and W. Karwowski (Eds.): AHFE 2019, AISC 960, pp. 122–130, 2020.
https://doi.org/10.1007/978-3-030-20488-4_12

testing phase, with an accuracy rate greater than 90% and a false positive rate of less than 4%. This shows that our model is effective and adaptable to the dynamic behavior of the user. On the other hand, a variability in the execution of user commands has been found to be quite high in periods of short time, but the proposed algorithm tends to adapt quite optimally.

Keywords: Intrusion detection · Digital fingerprint · Commands · Sequences · Pagerank

1 Introduction

The management of information in computer systems throughout the planet is an area of research with great potential, because cyber-attacks have increased exponentially in recent years. The battle to safeguard computer systems from new types of threats has forced experts to venture into multiple areas of knowledge such as artificial intelligence and mathematics. One of the greatest dangers faced by computer systems is the access of reserved information from internal and external intruders. These intruders impersonate other users and carry out fraudulent activities that harm in a financial way the institutions or people. There are multiple studies in that regard, where supervised and unsupervised learning have been applied from the field of machine learning and pattern recognition to increase the accuracy of identification within intrusion detection systems (IDS).

One of these studies is carried out by [2] where a new semi-supervised learning perspective is proposed. This proposal is based on confusion through the use of unlabeled samples, generated by means of a supervised learning algorithm to improve the efficiency of the classifier for the IDS. The detector uses a hidden neural network (SLFN) which allows to identify by means of categories the intrusive user. The data set used in this article is the well-known NSL-KDD, the same one that is based on the KDD99 of intrusive network information. The results obtained have been optimal and a comparison has been generated with other classification techniques already applied.

Another article about IDS is the one published by [3], where they propose a novel hybrid method of intrusion detection. This method integrates hierarchically a misuse detection model and an anomaly detection model in a breakdown structure. The study generates a misuse model by applying decision trees and later in smaller sets implements an anomaly detection model that uses attack information and user behavior profiles, resulting in a more accurate and efficient intrusion detection system. This study uses the NSL KDD database to train the classification algorithms. The detection time is significantly reduced, which benefits in an implementation in a real system.

In the article conducted by [4], the application of Genetic Fuzzy Systems in a pairwise learning is proposed, based on fuzzy sets as linguistic labels. This allows to determine more efficiently the set of fuzzy rules. On the other hand, the scheme of dividing is to win is applied, that is, it is compared with all possible pairs of classes with objectives. In this way it improves the accuracy of rare attack events by a large percentage. The results of this proposal were compared with that obtained with the decision tree technique C4.5.

Currently, several artificial intelligence techniques are applied, such as the k-nearest neighbors (KNN), as detailed in their work [5]. Their proposal is a representation of novel characteristics identifying the center of the data groups (cluster). This model measures and adds two distances, where the first distance is a sample and its center are the cluster of data. The second distance is determined between the information and its nearest neighbor in the same cluster.

A function that defines each data of the cluster is generated and later is entered into a KNN algorithm to obtain an intrusion detection with a quite optimal result. This proposal obtains a fairly high precision and a very low false positive rate.

Another approach in the detection of intruders is the one published by [6], the same that develops a new hybrid model that consists of the algorithms J48, Meta Pagging, RandomTree, REPTree, AdaBoostM1, DecisionStump and NaiveBayes. The objective of the study is to estimate the degree of intrusion scope in a computer network, identifying specific characteristics for the study. The experimental results obtained in this article showed that the proposed model minimizes computational complexity and detection time. In addition, the detection rate is close to 99.81%.

A study on intrusion detection is the one carried out by [7], which proposes an algorithm to detect anomalous behaviors of users in computer sessions. This model allows to detect a behavior profile of each user and identifies small anomalies in sequences of 2 commands within a user session. This model applies an algorithm based on the probability of these sequences. The activities classified as possible anomalies are verified by the application of Markov chains. This method has shown to be efficient in terms of high detection accuracy and low false positive rate.

In this paper we propose to use a command execution database in a UNIX operating system, where there are eight different users. With this information, a behavior profile of the execution of each user is generated and the activities that are most executed are determined. The execution probability of each command is calculated with the Page Rank algorithm [1, 8], which has been the fundamental basis for this study.

The work is structured as follows: In Sect. 2 the methods and materials used in the research are presented. The methodology used for the development of the intrusion detection model using data mining and the Page Rank algorithm is described. Section 3 details the proposed mathematical model and raises the phase of training and testing. Section 4 presents the results of the model with its accuracy and performance. In Sect. 5, a discussion of the results is presented comparing with works already published. Finally, we present the conclusions obtained in the research and future works.

The following section details the techniques used to analyze the information and the algorithms that formed the basis of the development of the intrusion detection model.

2 Materials and Methods

This section presents the database that was used to generate the intrusion detection model. On the other hand, the techniques of artificial intelligence to perform data mining are detailed, as well as the Page Rank algorithm which enables us to define the profiles of each of the users.

2.1 Materials

This section details the information that was used as a basis for the development of the model, such as the training and testing of the proposal.

This database contains a set of eight users, which were collected from computers with a UNIX operating system. This information belongs to the history commands execution during a period of two years. This data set is sequential, that is, each command was executed consecutively during the use of the computer. In addition, this database has flags that identify the start of each of the Shell sessions of execution of commands, using the tag ** SOF ** and ** EOF ** to determine the end of the session. The sessions of each user are within a text file arranged by date [9]. An example of the database is showed in Fig. 1.

```
**SOF**
cd
<1>                  # one "file name" argument
ls
-laF
|
more
cat
<3>                  # three "file" arguments
>
<1>
exit
**EOF**
```

Fig. 1. Sample of the UNIX database.

The distribution of information from the Unix database is shown in Table 1, where the length of the execution of commands in a session can be evidenced, grouped by users. In addition, the number of commands used in the entire database can be observed.

Table 1. Distribution of data per users in the UNIX database.

Users	Sessions	Commands	Maximum session length
USR_0	567	7840	214
USR_1	515	18851	429
USR_2	1069	16600	353
USR_3	501	15864	1865
USR_4	955	35905	957
USR_5	582	33657	1400
USR_6	3419	57314	434
USR_7	1522	14285	446
USR_8	1983	50076	564
TOTAL	*11113*	*250392*	

The data structure with which we are going to work is shown in Fig. 2, where a user session contains 1 or n number of commands executed. This describes a dynamic behavior on the use of the operating system.

```
    ** SOF                                                          ** EOF
**           f      <1>      f        <1>      f        <1>     **
    ** SOF                                                  ** EOF
**           elm    telnet   <1>      f        <1>      **
    ** SOF                                                          ** EOF
**           cd     <1>      cd       <1>      ls       cd      **
    ** SOF   teln                     ** EOF
**         et       <1>      lo    **
    ** SOF                                                          ** EOF
**           f      <1>      date     telnet   <1>      f       **
    ** SOF                                                  ** EOF
**           f      <1>      telnet   <1>      elm      **
    ** SOF                                                          ** EOF
**           elm    vt100    elm      telnet   <1>      telnet  **
    ** SOF                                                  ** EOF
**           f      <1>      date     elm      ls       **
    ** SOF                                                          ** EOF
**           elm    f        <1>      date     elm      fg      **
    ** SOF                                     teln                 ** EOF
**           elm    f        <1>      date   et         <1>     **
    ** SOF          ** EOF
**           lo   **
    ** SOF                   ** EOF
**           elm    lo     **
```

Fig. 2. Sequential data structure of the UNIX database.

A user session is defined as S_m^u, where m is the number of sessions that contains the history of each user u. Each session S contains 1 or n number of commands executed, defined as C_n, where n is the position in which the command has been executed. A user session would be defined as $S_m^u = \{C_1, C_2, C_3, \ldots, C_n\}$. Within each user session S_m^u there are several subsequences of two commands, and these subsequences are defined as $sb = \{C_n, C_{n+1}\}$, which provide information on which command C_{n+1} is executed after the C_n command.

2.2 Methods

This section presents the data mining algorithms that are used to pre-process the information in the UNIX database, as well as the PageRank algorithm which allows defining the profiles of each user.

Instance and attribute selection algorithms

Multiple data mining techniques are necessary to perform the pre-processing of data. These techniques allow us to identify which instances and attributes are correct and provide a large amount of information to the study.

Greedy Stepwise Algorithm

In data mining, greedy algorithms are used, which are a search that applies a heuristic to identify the locally optimal choice to obtain a globally optimal solution. These algorithms are used in optimization problems.

The mathematical definition is based on a finite set of C inputs, a greedy algorithm returns a set S (selected data) such that $S \subseteq C$ and that also complies with the constraints of the initial problem. For each set S must comply with the restrictions, in addition, must achieve that the objective function is minimized or maximized (depending on the case), resulting in a set S of optimal solutions [10].

Automatic Noise Reduction

During the data collecting and labeling process, noise can usually be found in a data set. Due to this, the quality of the data is low and the classification or detection models are less reliable. The algorithm of Automatic Noise Reduction (ANR) is used as a filtering mechanism to identify elements with noise or that have been mislabeled. This mechanism applies the multi-layer neural networks, which assign to each element of the data set a provisional label, which may vary during the training of the algorithm. Subsequently, when the noise percentage is less than a threshold of 30% of the data, the data elements whose classes have been re-labeled are processed as noisy data and are removed from the data set [11].

PageRank

The PageRank algorithm is a method to list web pages objectively and mechanically, effectively measuring the human interest and attention devoted to them. This method measures the number of calls made from an initial web page to another determined web page. Subsequently, a visit probability of each web page is obtained and it is listed from highest to lowest probability, facilitating the search of the user [8]. This algorithm describes an approximation of the user's navigation in real time [1]. The ranking of the pages is defined by Eq. (1).

$$R'(u) = c\sum\nolimits_{v \in B_u} \frac{R'(v)}{N_v} + cE(u) \tag{1}$$

The following section presents the development of the proposed model using the Unix database as a fundamental resource for the generation of user profiles and the PageRank algorithm to identify the probability of execution of commands.

3 Development of the Intrusion Detection Model

In this section we present the intrusion detection model by applying PageRank to the database. As a first phase, a pre-processing of data must be performed in which the instances with noise must be eliminated by applying the Automatic Noise Reduction algorithm. The result of the elimination of 3% of data with noise in the information of the 8 users has allowed a more optimal database.

The Greedy Stepwise algorithm has been applied to identify the most important commands, of which 17 commands has been identified as the most executed by the users. With this database, 43% of the information that was not relevant to the study has been eliminated.

To identify the probability of the most used commands by each user, it is necessary to make a square matrix $M_{n,n}$, which contains 17 columns and 17 rows. The matrix $M_{n,n}$ will be described by each command C_n, where if a command C_{n+1} is executed after a command C_n in the matrix $M_{n,n+1}$ it will be assigned the value of 1, otherwise the value of 0. An example is presented in the following matrix, where we only work with 5 commands ($n = 5$):

$$M_{5,5} = \begin{bmatrix} 0 & 0 & 1 & 1 & 1 \\ 0 & 0 & 0 & 0 & 1 \\ 1 & 0 & 0 & 1 & 1 \\ 1 & 0 & 1 & 0 & 0 \\ 1 & 1 & 0 & 0 & 0 \end{bmatrix}$$

Afterwards, an addition will be made per column of the matrix M, defined as $sum_i = \sum_{i=1}^{n} M_{n,i}$, and then all the values of each column will be divided by sum_i, obtaining the following matrix M' :

$$M' = \begin{bmatrix} 0 & 0 & 1/2 & 1/2 & 1/3 \\ 0 & 0 & 0 & 0 & 1/3 \\ 1/3 & 0 & 0 & 1/2 & 1/3 \\ 1/3 & 0 & 1/2 & 0 & 0 \\ 1/3 & 1 & 0 & 0 & 0 \end{bmatrix}$$

The next step is to calculate the v value of the pagerank, as shown in Eq. (1). We must take this matrix M' to a system of auto vectors defined as $M'v = \lambda I_d v$, where $\lambda \in \mathbb{R}$. We add up the additive inverse to the equation obtaining a homogeneous system and calculating the value of pagerank v, as shown:

$$(M' - \lambda I_d)v = 0$$

$$\left(\begin{bmatrix} 0 & 0 & \frac{1}{2} & \frac{1}{2} & \frac{1}{3} \\ 0 & 0 & 0 & 0 & \frac{1}{3} \\ \frac{1}{3} & 0 & 0 & \frac{1}{2} & \frac{1}{3} \\ \frac{1}{3} & 0 & \frac{1}{2} & 0 & 0 \\ \frac{1}{3} & 1 & 0 & 0 & 0 \end{bmatrix} - \begin{bmatrix} \lambda & 0 & 0 & 0 & 0 \\ 0 & \lambda & 0 & 0 & 0 \\ 0 & 0 & \lambda & 0 & 0 \\ 0 & 0 & 0 & \lambda & 0 \\ 0 & 0 & 0 & 0 & \lambda \end{bmatrix} \right) \begin{pmatrix} C_1 \\ C_2 \\ C_3 \\ C_4 \\ C_5 \end{pmatrix} = 0$$

Where the value v of each of the commands C_n is calculated, obtaining the following result $C_1 = 6$, $C_2 = 1$, $C_3 = 5.33$, $C_4 = 4.66$ and $C_5 = 3$. Obtaining the probability of each command $P(C_1) = 1$, $P(C_2) = 0.16$, $P(C_3) = 0.88$, $P(C_4) = 0.77$ and $P(C_5) = 0.50$. With these results, we can see which command is the most and least used by the user.

For the Unix database, this process has been carried out with the eight users, where the most commonly used commands by each user have been detected. To identify the normal behaviors, it has been proposed to verify if the commands per each user have a probability of being executed in new sessions. In the event that the probability is low or non-existent, a warning message of “possible intruder” will be launched.

The experimental results of the intrusion detection model are presented in the next section. In addition, the results obtained with other previously published works will be discussed.

4 Results

The results obtained in this study have been quite optimal, the tasks most executed by the user have been identified in a more precise way. In this way an anomalous and possibly intrusive behavior can be identified. Table 2 shows the results of detection accuracy in both algorithm training and testing.

Table 2. Results of the training and tests of the algorithm proposed with the UNIX database.

	Training		Test	
User	Accuracy	Error	Accuracy	Error
USER0	89,71	10,29	89,00	11,00
USER1	88,56	11,44	90,21	9,79
USER2	88,89	11,11	88,12	11,88
USER3	90,01	9,99	90,45	9,55
USER4	89,46	10,54	89,31	10,69
USER5	88,00	12,00	88,02	11,98
USER6	87,88	12,12	87,32	12,68
USER7	90,12	9,88	90,05	9,95
Average	89,08	10,92	89,06	10,94

With these results it has been shown that the accuracy rate of this algorithm is optimal with an average of 89.08% in training and 89.06% in the testing phase, compared with the results obtained by [7], of correctly classified of 94.78%.

The conclusions obtained in the development of this research are presented in the following section, as well as the future lines of research that emerged during the study.

5 Conclusions and Future Works

In this work we have identified points of great importance to take into consideration, because they provide much information to efficiently detect an intrusive behavior. As a fundamental basis, the execution of tasks in a sequential manner is essential to develop a profile of a user’s behavior in a computer system, due to in small sequences the human being repeats activities constantly during the use in a system. This study showed that there are commands that are the most and least used by each user, and also vary significantly depending on the position and responsibilities. The application of the pagerank algorithm allows us to design a dynamic profile that evolves over time.

This is a great advantage, since a user is changing constantly their way of interacting and working. The results obtained are optimal, but they are still a theoretical approach to design a complete and efficient user profile. As future research lines, we intend to apply hidden markov networks to accurately determine an intrusive behavior. In addition, multiple generic user profiles will be generated according to position and responsibilities, which will allow a possible intrusion to be detected in a more real way.

References

1. Gleich, D.F.: PageRank beyond the web. SIAM Rev. **57**(3), 321–363 (2015)
2. Ashfaq, R.A.R., Wang, X.-Z., Huang, J.Z., Abbas, H., He, Y.-L.: Fuzziness based semi-supervised learning approach for intrusion detection system. Inf. Sci. (Ny) **378**, 484–497 (2017)
3. Kim, G., Lee, S., Kim, S.: A novel hybrid intrusion detection method integrating anomaly detection with misuse detection. Expert Syst. Appl. **41**(4), 1690–1700 (2014)
4. Elhag, S., Fernández, A., Bawakid, A., Alshomrani, S., Herrera, F.: On the combination of genetic fuzzy systems and pairwise learning for improving detection rates on intrusion detection systems. Expert Syst. Appl. **42**(1), 193–202 (2015)
5. Lin, W.-C., Ke, S.-W., Tsai, C.-F.: CANN: an intrusion detection system based on combining cluster centers and nearest neighbors. Knowl.-Based Syst. **78**, 13–21 (2015)
6. Aljawarneh, S., Aldwairi, M., Yassein, M.B.: Anomaly-based intrusion detection system through feature selection analysis and building hybrid efficient model. J. Comput. Sci. **25**, 152–160 (2018)
7. Guevara, C., Santos, M., López, V.: Data leakage detection algorithm based on task sequences and probabilities. Knowl.-Based Syst. **120**, 236–246 (2017)
8. Page, L., Brin, S., Motwani, R., Winograd, T.: The PageRank citation ranking: bringing order to the web, November 1999
9. Aeberhard, S., Coomans, D., Vel, D.: UCI Machine Learning Repository: UNIX User Data Data Set. https://archive.ics.uci.edu/ml/datasets/UNIX+User+Data. Accessed 17 Dec 2018
10. Zarkami, R., Moradi, M., Pasvisheh, R.S., Bani, A., Abbasi, K.: Input variable selection with greedy stepwise search algorithm for analysing the probability of fish occurrence: a case study for Alburnoides mossulensis in the Gamasiab River, Iran. Ecol. Eng. **118**, 104–110 (2018)
11. Xinchuan, Z., Martinez, T.: A noise filtering method using neural networks. In: IEEE International Workshop on Soft Computing Techniques in Instrumentation, Measurement and Related Applications, SCIMA 2003, pp. 26–31 (2003)

Investigation and User's Web Search Skill Evaluation for Eye and Mouse Movement in Phishing of Short Message

Takeshi Matsuda[1(✉)], Ryutaro Ushigome[2], Michio Sonoda[3], Hironobu Satoh[3], Tomohiro Hanada[3], Nobuhiro Kanahama[3], Masashi Eto[3], Hiroki Ishikawa[3], Katsumi Ikeda[3], and Daiki Katoh[3]

[1] University of Nagasaki, Nishi-Sonogi-gun, Nagasaki, Japan
tmatsuda@sun.ac.jp
[2] Chuo University, Bunkyo-ku, Tokyo, Japan
[3] National Institute of Communications and Technology, Koganei, Tokyo, Japan
{sonodam, satoh.hironobu, hanada, kanahama, eto, ishikawa, ikeda, dkatoh}@nict.go.jp

Abstract. There are many studies on eye and mouse movement. However, there are not many studies that try to evaluate the skill of Web search while considering the relationship between the line of sight and the movement of the mouse. In this study, we analyze the data acquired from the viewpoint of the differences in information literacy of subjects and investigate the method of quantitatively evaluating the skill of web search.

Keywords: Eye and mouse movement data · Web search skill · Skill evaluation

1 Introduction

With the development of ICT technology and the spread of smart devices, various kinds of information are accumulated on the Internet. Big data analysis is very important technology of finding valuable information from such a large amount of data, and research is continuing in various fields. Although development of such big data analysis technology can be said to be an important social theme, it can be said that the web search capability that enables users to quickly find the necessary information from the Internet is also an important skill. Ease of finding information on Web pages can be quantitatively evaluated by Fitz's law [1]. Fitz's law can also be used to improve web pages to make it easier to find information through web design. However, the information required by individual users may be included in web pages that do not consider UI. Also, because there are some information that are hard to get caught in the top ranking in web search, skill related to web search is important in order to obtain appropriate information quickly.

The purpose of this study is to investigate how user's web search skills can be reflected in data by collecting gaze and mouse data during web search.

T. Ahram and W. Karwowski (Eds.): AHFE 2019, AISC 960, pp. 131–136, 2020.
https://doi.org/10.1007/978-3-030-20488-4_13

It is well known that the gaze and mouse movement data has correlation [2]. This study had investigated the process of web search using the theme that is difficult to hit by web search. These days, the phishing method using SMS is becoming a social problem. It is not easy to find the evidence of SMS phishing by web search. So, we had treated the theme of information search on phishing using SMS.

Some SMS phishing is observed when an emergency such as an earthquake or a heavy rain. Because the amount of information sent by SMS is small, at first glance there is important information on emergency situations and the link URL. Since SMS can write only a small amount of information, it can be said that it is a natural way to have additional information refer to URL of link. Users who are familiar with phishing techniques will check to see if other users are providing information on the Internet or not by without clicking on that URL. So, by investigating the process of how to perform web searching, we will study what kind of difference can be seen between people with high IT literacy and people with low IT literacy.

2 Related Study

In this research, we had collected the data on gaze and mouse movement during web searching, and investigate the characteristics of those data. Conventional research on gaze and mouse movement and conventional research on Web search will be introduced in this section.

2.1 Study on Gaze and Mouse Movement

The paper [1] had pointed out that the gaze and mouse movement has strong correlation. Collecting data is not easy because gathering gaze data requires professional equipment such as eye tracker. On the other hand, it is not difficult to collect the coordinate data of the mouse on the display, and it can be said that data collection is easier compared to the gaze movement data. Software that suggests a method of improving web page design and UI from mouse movement based on application of correlation between eye and mouse is commercialized [3]. In addition, applications that use gaze instead of mouse operation have also been developed [4].

2.2 Research on Web Search

The paper [5] had considered estimating the gaze movement from the mouse movement during web search. They had concluded that it is not appropriate to estimate the gaze movement using the mouse movement at every action during the web search. There are also studies that detect obstacles related to networks from users' web search and posting by SNS [6].

Equations should be punctuated in the same way as ordinary text but with a small space before the end punctuation mark.

3 Experiment

We did 2 types of experiments as the below in order to collect the data of eye and mouse movement when a user searches information on the web.

Experiment 1

Each subject was shown the screenshot of SMS whose content is phishing, and we examined that whether he/she found the screenshot the phishing message or not (Fig. 1).

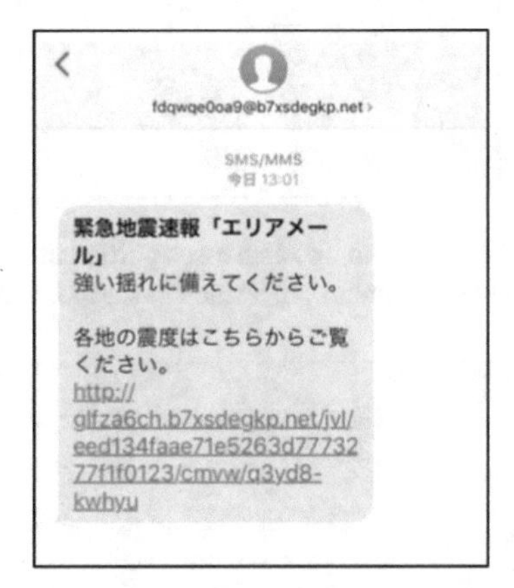

Fig. 1. Screenshot of phishing SMS

Experiment 2

After the Experiment 1, each subjects searched the evidence indicating that the SMS is a phishing message. All subjects used Google or Bing as the search engine.

[Environment]

Eye-tracker: Tobii X2-30 Compact Mouse Cursor Tracing: UWSC OS: Windows10 Pro 64bit Number of Subjects: 4

4 Result and Consideration

4.1 Analysis Using AoI

We examined which part of the SMS image the subjects focused on using AoI (Area of Interest). The areas we focused in SMS are shown in Fig. 4 and the result of AoI is shown in Fig. 4. In Fig. 4, the vertical axis indicates the length of time for which each subject gazed in the areas. We can see from Fig. 4 that Subject 1 and 4 gazed the source address most, Subject 2 gazed the title of the message most and Subject 3 gazed the URL most which leads the user to a phishing site (Figs. 2 and 3).

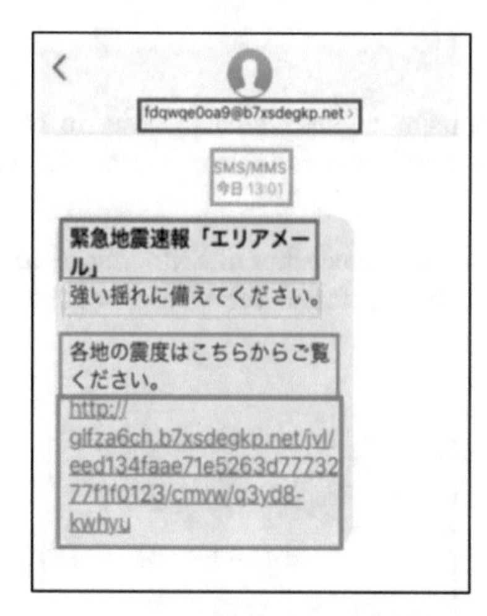

Fig. 2. AoI on the phishing SMS. The source address is written in the blue area, the received time is written in the orange area and the contents are written in the red, yellow sky blue and green area.

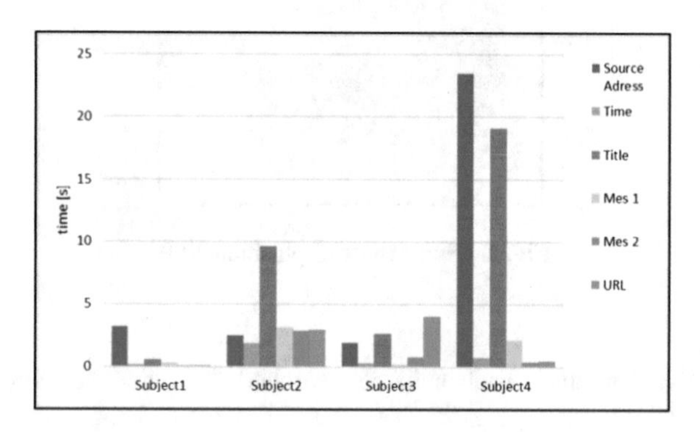

Fig. 3. The result of AoI

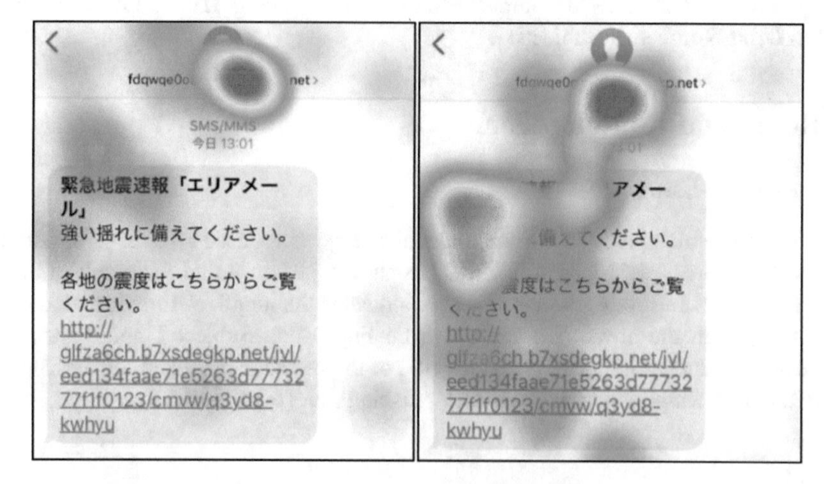

Fig. 4. Heatmaps of Subject 1 (left) and Subject 2 (right)

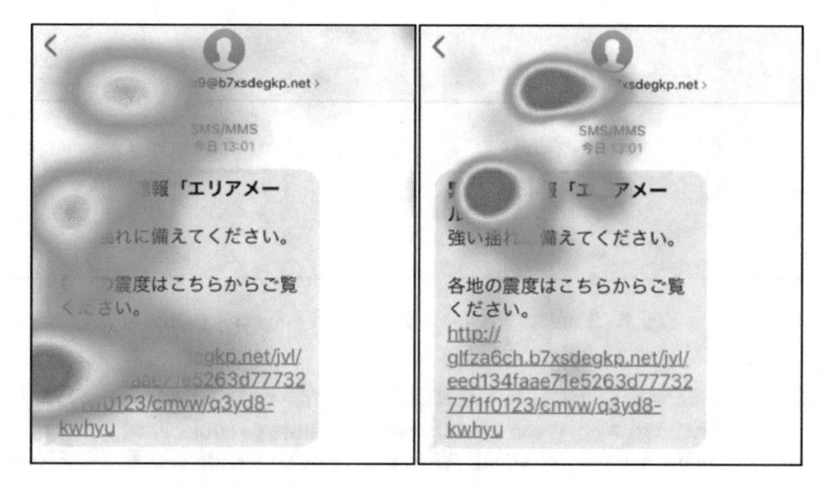

Fig. 5. Heatmaps of Subject 3 (left) and Subject 4 (right)

All subjects had focused attention on the word "Emergency Earthquake Warning", in common. It can be seen that subjects with high literacy gaze at the domain part of the e-mail address. Because of the small amount of data, there are no major features of AoI other than gaze to the domain. However, it can be said that the result that we cannot deny the possibility of evaluating the presence or absence of literacy from the place the subjects gaze eye movement is obtained.

4.2 Analysis of Eye and Mouse Movement Data

We examined the correlation between eye-tracking and the motion of mouse cursor with the analyzing method proposed in [7] for understanding the relationship between them. Figures 5 and 6 show the scatters of the data obtained through the experiments stated in the prior chapter. The horizonal axes in each graph indicate the motion distance of the mouse cursor and the vertical axes indicate the motion distance of eye gaze (Fig. 7).

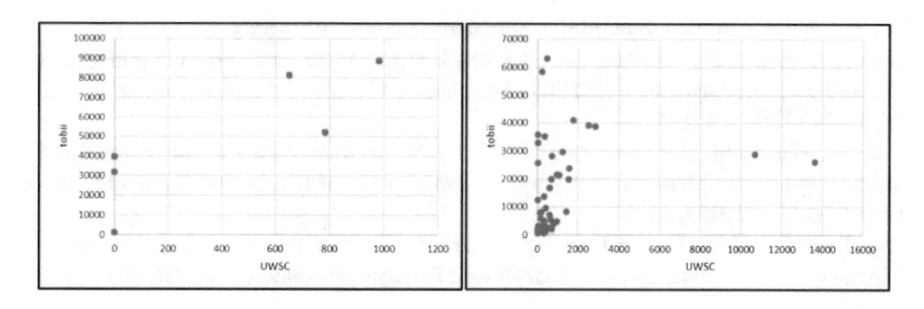

Fig. 6. Scatters of Subject 1 (left) and Subject 2 (right)

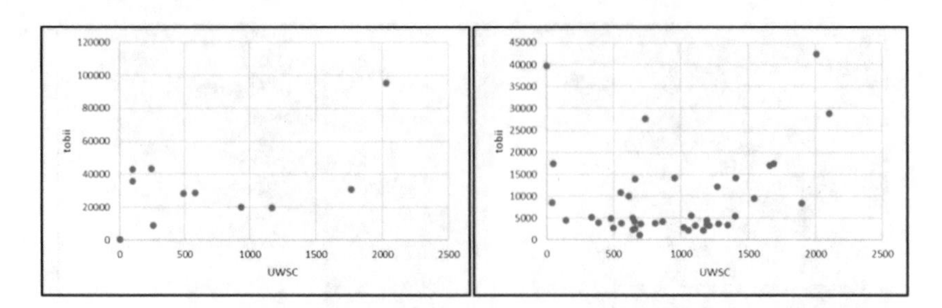

Fig. 7. Scatters of Subject 3 (left) and Subject 4 (right)

What is common to all subjects is that the mouse movement distance has little effect on literacy differences. And, it can be said that the feature of subjects with low literacy, the eye movement distance is small. Therefore, subjects with low literacy have no correlation between eye and mouse movements. It can be said that the characteristic of presence or absence of literacy like this has regression line equation.

5 Conclusion

In this study, we had analyzed user's eye and mouse movement data during web search on SMS phishing, and examined its features. However, since the number of subjects is small, it is the our future work to investigate whether the same result can be obtained even if the number of subjects is increased.

References

1. Fitts, P.M.: The information capacity of the human motor system in controlling the amplitude of movement. J. Exp. Psychol. **74**, 381–391 (1954)
2. Chen, M.C., Anderson, J.R., Sohn, M.H.: What can a mouse cursor tell us more?: correlation of eye/mouse movements on web browsing. In: Proceedings of CHI 2001, pp. 281–282. ACM (2001)
3. https://sitest.jp
4. https://www.microsoft.com/ja-jp/enable/products/windows10/eye.aspx
5. Huang, J., White, R., Buscher, G.: User see, user point: gaze and cursor alignment in web search. In: Proceedings of the SIGCHI Conference on Human Factors in Computing Systems, pp. 1341–1350 (2012)
6. Oki, M., Takeuchi, K., Uematsu, Y., Ueda, N.: Mobile network failure detection and forecasting with multiple user bahavior. In: Proceedings of the Annual Conference of JSAI, JSAI2018(0), 1D105 (2018)
7. Web Search Skill Evaluation from Eye and Mouse Momentum. In: Proceedings of the 2nd International Conference on Intelligent Human Systems Integration (IHSI 2019), pp. 884–889 (2019)

Author Index

T. Ahram and W. Karwowski (Eds.): AHFE 2019, AISC 960, pp. 137–138, 2020.
https://doi.org/10.1007/978-3-030-20488-4

Printed in the United States
By Bookmasters